Social Manipulation

By

Roderick Larios

This book is a work of fiction. Places, events, and situations in this story are purely fictional. Any resemblance to actual persons, living or dead, is coincidental.

© 2003 by Roderick Larios. All rights reserved.

No part of this book may be reproduced, stored in a retrieval system, or transmitted by any means, electronic, mechanical, photocopying, recording, or otherwise, without written permission from the author.

ISBN: 1-4107-8577-7 (e-book)
ISBN: 1-4107-8997-7 (Paperback)
ISBN: 1-4107-8996-9 (Dust Jacket)

Library of Congress Control Number: 2003095517

This book is printed on acid free paper.

Printed in the United States of America
Bloomington, IN

1stBooks – rev. 10/03/03

A special thanks to Jeremy Longstreet for the outstanding artwork on the cover. Jeremy's still life's can be viewed at (www.Jeremylongstreet.com)

Also I'd like to thank the following people who actually took the time to read the original manuscript and gave me both the positive and negative feedback I needed to complete the book.

Lance and Kathy Hubener
Janice York
Kay Lang
Kathy Mercatante
Michelle, Dominic and Margaret Jacobellis
Jennifer Bell
Gina Bennett
Dee and Wrenda Richardson
Mona Oneal
And Becki and RJ

Chapter I

Aaron Douglas could see brook trout wiggle their way over the shallow gravel beds in Sutter's Creek. It was 5:30 A.M. on a clear May morning. The sun was shedding the first rays of brilliant light through the tall Ponderosa Pines. Aaron's scouting trips were his passion. Though he loved to hunt, the scouting was what he lived for. He'd always pack the same gear, a backpack with food, a rope, a canteen of water, field glasses, a knife, and his twenty-two-caliber pistol. He would have carried a map and compass, but his land navigation skills left something to be desired, so he stayed on the south side of the mountain range, knowing that by merely walking downhill he would always run into the Little Brazier River.

Less than an hour into his hike, Aaron came across a fresh set of elk tracks, a cow and calf as near as he could tell. By this time he was about a quarter way up the mountain. He decided to follow the tracks. Through a stand of Aspen, he caught a glimpse of the two elk. He was just short of the trees when he heard the cow bawling; she couldn't have been more than 200 yards off by the sound. He'd rarely heard elk be so vocal and wondered what was happening. As he reached the far edge of the Aspens, he saw a sight that made the hair on the back of his neck stand up straight. The biggest black bear he'd ever seen was running full speed through the trees toward him, with the elk calf in its mouth.

The bear wasn't more than 50 yards away, and Aaron was almost in shock. All he could do was watch in horror as the bear ran by him, a mere 10 yards away, and headed down a draw with the calf in its mouth and the cow in tow. All the commotion was over in a few seconds, and a more nervous Aaron continued on the trek up the mountain. He had heard about how aggressive black bears were during calving season, but had never witnessed an event such as this. He felt sorry for the cow and calf. He thought about how foolish it was to prohibit bear hunting with hounds. If people could have witnessed this event, they may have formed different opinions about fuzzy little bears.

About an hour after the bear event, Aaron was over halfway up the mountain. He spotted four nice bucks; their velvet antlers glistened in the sunlight. Though their antlers weren't fully developed, he could tell they were massive. He watched them for a few minutes and started back up the mountain. He sat on a large boulder overlooking a stand of Lodge Pole Pine and ate a sandwich and an apple. About a quarter mile across a draw, he thought he saw movement and quickly put his field glasses on the sight. By the time he zeroed in on the spot, there was nothing moving. Then a reflection and a faint echo drifted up the draw. This time he could make out a girl hammering something into a tall Ponderosa Pine tree.

He moved slowly toward the girl and when he was within 50 feet he yelled, "What in the hell do think you're doing?" She was so shocked that she dropped the hammer and screamed at the top of her lungs. She was shaking when he reached her, and he picked up the hammer and handed it to her. He asked her the question again. "What in the hell do think you're doing?" Her answer didn't surprise him.

"I'm saving the forest from the loggers," she retorted. He was shaking his head in amazement.

You're not saving anything, you idiot, he thought to himself, but was less aggressive when he spoke. "Why is driving a spike into a tree going to save the forest?" Her explanation was the usual environmentalist lies about how logging hurts the forest, and all the animals and fish and air and everything else that was in or near the forest. "Do you really think that cutting down trees is going to ruin the forest?" Aaron asked.

"Yes," she replied.

He pointed out the fact that the very area they were in had been logged before and yet was still a beautiful forest. He asked her if she really knew anything about forestry, and she didn't. She was just another idealistic college kid out to save the world. "What's your name?" Aaron asked.

She wouldn't tell him, but he volunteered his. "My name is Aaron. Do you know how much trouble you're in?" She started to cry, and he had to laugh.

"What do you find so funny?" She sobbed. He told her that a real terrorist wouldn't think about crying. They would just take their punishment, and be proud of what they had done.

He didn't like what she was doing, but there was something about her that was appealing to him. He found out that she was up on the mountain with a friend, another girl out to save the forest. He told her about the bear and suggested they find her friend before something else found her. The girl agreed, and they started back down the mountain.

They found her friend, fresh out of spikes and shaking like a leaf. "My God," she said. "I just saw the worst thing I've ever seen in my life."

"Anything to do with a bear?" Aaron asked.

"Yes! The bear was eating this big deer's baby, and the mother deer tried to save it, but the bear just tried to catch her too. It was awful." She went on to explain how she had just finished spiking her last tree when she heard the commotion and went to investigate. She was pretty shaken, and Aaron suggested they get moving.

This second girl was short, about 5'2" or so, and wore Nike shoes, Levis, and a U of O sweatshirt. The first girl was taller, with light brown hair, green eyes, and a few freckles. She wore plain white tennis shoes, jeans, and a windbreaker. He thought she was good-looking, in a wholesome sort of way. As they walked down the mountain, she asked why he had been up the mountain so early. Aaron explained to her about the scouting trips he took from time to time, and then told her he'd have to turn them both in to the county sheriff for spiking the trees.

"What can they do to us? We didn't hurt anybody," the short girl asked. Aaron explained to her that a couple of years back, two "tree-savers" had been caught and done some jail time, as well as community service.

About halfway down the mountain, they came across an old logging road. The tall girl said that their car was just down the road and they would be okay. "How did you get a rig up here?" Aaron asked. "That gate has been locked up for the last three years." She said it didn't really matter, and he could just leave them now. Aaron explained to her that he wasn't going anywhere until he got a license number. The short girl began to cry and mildly pleaded with him to forget he'd ever seen them.

"Do you guys know what happens when a 12-foot band saw hits one of those spikes?" Aaron asked. Without a pause, he answered his own question. "They fly to pieces. My dad says they are worse than a hand grenade when they hit just right. Hand grenades kill people, and so do those spikes you're nailing in the trees."

"So what do you plan on doing to us?" the tall girl asked.

"Nothing," Aaron said. "The courts will decide what to do." No one said a word for a minute or two. There was a warm breeze, and they could hear the wind blowing through the trees. Aaron spoke first. "I don't like to see anybody go to jail, especially over here, the people on the east side of the state take things like this seriously." The tall girl asked if promising never to do anything like that again was good enough. Aaron thought for a moment, and then spoke. "What if we make a deal? You keep an open mind about the lumber industry, and I'll keep an open mind about the environmental terror business. In other words, you come into my world for a while, and I'll get into your world for a while, and we'll all go up and try to pull as many spikes as we can find out of the trees." Both girls reluctantly agreed, thinking it was the better of the two options. "So what are your names?"

"I'm Rachael Mayfield," the short one said." "And this is Emily Alcott, like the writer. What's your name?" Rachael asked. "I'm Aaron Douglas, like the fir tree." No one laughed, and Aaron figured neither of the girls knew much about trees.

When they reached the girls' Ford Explorer, Emily asked Aaron if he wanted a ride to his car, which he accepted. As they passed through the gate at the bottom of the logging road, Aaron asked again, "How did you get a key to this gate?" Neither girl answered, and there was a moment of silence.

"Look," Emily said. "Let's just stick to our deal, and if and when you come to understand why we were up here doing this, we will fill you in on our little secrets."

"Fair enough," Aaron said.

When they got to Aaron's pickup, they dropped him off and followed him into LaGrande. They decided that they would go back up and get out all the spikes they could that afternoon. They dropped the Explorer off at the motel where the girls were staying, and took Aaron's pickup. They headed south, and about 10 miles out of town, came to a large complex, which the girls recognized as a tree slaughterhouse. Aaron, of course, thought of it as the lifeblood of the entire area. The sign read "Douglas Land and Timber, The Largest Employer in the County." "Nice name," Emily said. "Any relation?"

"Yeah," Aaron said proudly. "My dad and the bank own this place." The girls waited while Aaron went into the maintenance shop and got a gadget neither of the girls had ever seen. Aaron explained that when a spike or nail was detected with the metal scanner, they used the large nail puller to retrieve them. Both girls actually seemed interested.

Rachael asked, "If they get the spikes out of the logs when they get here, why is it such a big deal that we were putting them in the trees?"

"Can't find all of them, even with the most sophisticated equipment," Aaron explained. They went to Aaron's folks' house, which was a couple of miles up a draw from the mill. Aaron ran in and came out with his 30-06 rifle slung over his shoulder.

"I hate guns," Emily confessed. "But with the stupid bear and all, I'll feel better if we have it with us." Aaron slid four shells into the magazine and left the bolt open. He placed the gun muzzle down beside him in the seat, and Emily got in the middle and Rachael on the passenger side. In less than half an hour, they were back at the scene of the crime, as Aaron referred to the place where they had spiked the trees. Aaron suggested that they stay together and start low, working up the hill to the north. It was just after two in the afternoon when they reached what they figured were the highest stand of Ponderosa that were affected. Emily actually knew where she was and where she had been.

"How do you know your way around so well?" Aaron asked. Emily went on to explain that her dad had been a platoon leader in Vietnam, and when she was growing up, they used to practice map and compass reading as a form of recreation. Aaron was extremely impressed. They found what they thought were most of the spikes just before dusk, and made it back to the truck just before dark.

"What now?" Rachael asked.

"Why don't we go to Sam's and get something to eat?" Aaron suggested.

Emily asked, "Do they have anything besides steak and potatoes?"

"Nope, that's it," Aaron replied.

Emily and Rachael both had fish and chips, and Aaron had a hamburger. It was Saturday night, and they decided that they would hash out the details of their agreement tomorrow. Aaron dropped them off at the motel and made plans to meet them in the morning. "Don't forget DDP 1TC," he said as he drove off.

"What's that?" both girls asked at the same time.

"Your license number," he explained.

Roderick Larios

Chapter II

    Emily and Rachael grew up in Tigard, a small town south of Portland. They had been best friends since third grade. Just finishing their third year at the U of O, they were still good friends, but had not seen that much of each other for the past two years. Emily was an education major, and Rachael was majoring in political science. Rachael had met Emily again at a sorority function in early January and told her about her involvement in an underground organization known as Friends of the Forest. She had convinced Emily to help her and her friends try and save the forest from the evils of mankind. At first, Emily had rejected the idea, but some of Rachael's friends were very convincing. A student by the name of Herb Snyder had recruited Emily over a period of three or four weeks. She finally gave in and became an official FF, as they referred to the organization in public. The tree spiking had been the second mission that Emily had been on. The first mission was a failed arson attempt on a ranger station on the Willamette Pass. They were convinced that what they were doing was justified and right. The leaders, whom Emily and Rachael had never seen, were very good at what they did. In the name of saving humanity, they set out to destroy any and all forest-related industry. They preached that their mission was to save the old-growth forest, but they targeted everything related to the timber industry.

    Rachael had been a member of FF for over two years, and had enough activity to land her some serious jail time. The only personal contact she had was with Herb Snyder, and another member by the name of James Sheldon. She had no idea where the orders came from, or how they were received. All she knew was that Herb called secret meetings prior to each mission, always in a different location. The last meeting she had attended was in a small tavern outside of Monmouth a week before the spiking was to take place. Only she and Herb were present. Herb had given her a map, spikes, hammers, a key to the forest service road, and 300 dollars in cash. The map was a forest service proposed timber sale map, showing the area that was coming up for bid in a couple of months. Their mission was to spike 40 trees, randomly spread out over 150 acres. Herb had told Rachael that the overall objective of FF was to get spikes into every timber sale area that was being proposed by the forest service. Each sale was then posted on a website, with pictures of the sale area map and pictures of the spiked trees. The FBI and other law enforcement agencies could not trace the origin of the pictures. They were mailed in plain, "no-return" envelopes to a non-FF member, who posted everything that was sent to him on the web. The state's Supreme Court, under the premise that the site was protected under the First Amendment, did not deem the activity illegal.

Rachael was to tell Emily only that the spiking was a random act, intended to harass local timber companies in the area. Rachael, however, had confided in Emily, and had told her everything that Herb had told her. Emily told Rachael that the group was more radical than she had first thought. Rachael had, at one time, tried to quit FF, and had been told by Herb that it was out of the question. He had explained that anyone trying to quit would have a file turned over to the FBI, relating each act of terrorism that he or she had committed. Rachael hadn't bothered to share that information with Emily. Emily had never said anything about wanting to quit FF, and Rachael hoped she never would.

They were forbidden to tell anyone anything about FF, without first getting permission from the higher-ups. Each candidate was first screened, and permission to proceed with the recruiting was either approved or denied. Even Herb had no idea who conducted the screenings. None of the rank and file knew much about the inner workings of FF. All of the court cases that had taken place were well defended, although some members had been given jail terms. Once a member was convicted, or charged with a crime, he or she was no longer a member. That way, only non-offenders were members. Rachael and Emily were both in over their heads, and neither of them really knew why they had joined. It just seemed right at the time.

Aaron Douglas grew up in Baker, where his dad, A.J. was superintendent of a good-sized sawmill. He was a good athlete, competing in football and wrestling. His mother had died in an automobile accident when he was four, and A.J. had raised Aaron and his younger sister, Mary, pretty much on his own. When Aaron was 14, and a freshman in high school, A.J. had remarried. At first, Aaron and Mary both hated Nancy, but over time, they had become very fond of her. She had always treated them as her own, and never had other children. Shortly after A.J. and Nancy were married, the nearly broke and very rundown Willow River Sawmill near LaGrande had come up for sale. With the mill, there were 47,000 acres of prime timberland. A.J. had been a good saver over the years, and had done well on some investments. Nancy had also saved a fair amount, and together they came up with enough to get in the door of what became Douglas Land and Timber. By the time Aaron graduated from high school, Douglas Land and Timber had become the largest employer in the county, with over 200 employees. They had their own logging company with nine trucks, and produced over 400,000 board feet a day of prime Ponderosa Pine factory grade and dimension lumber. A.J. and Nancy worked long hours to get the mill off the ground, and Aaron and Mary were on their own most of the time. A.J. and Nancy rarely missed one of Aaron's games or matches. They also made it to most of Mary's volleyball games and track meets. Both Aaron and Mary excelled in school. Nancy had been an elementary school teacher before they bought the mill and helped Aaron and Mary with their studies whenever she had a chance. Aaron had just completed his third year at OSU, majoring in forestry, and Mary was in her senior year of high school. Aaron had worked at the mill every summer since his junior year in high school and had loved every minute of it. He liked working with the foresters most of all, and knew that forestry was going to be his life.

Roderick Larios

Chapter III

It was 7 A.M. when Aaron pulled his F-150 into the parking lot of the Best Western. They had planned to meet at 7:15, and Aaron wondered if they would be there. The Explorer was parked in the lot where it had been the night before. Rachael came out first and smiled and waved. Emily was a few steps behind her, and when Aaron saw her his mouth fell open. She was wearing shorts and a halter-top and hardly looked like the tree-spiker from the day before. She was no longer pretty in a wholesome sort of way. In Aaron's mind, she was just plain beautiful. He wondered if the fresh look might be for his benefit. He happened to be wearing his best jeans and cowboy boots and his favorite eight X Stetson hat. Emily stood and looked at Aaron and his 6'2", 200-pound frame, and started to laugh.

"What's the matter?" Aaron asked.

"Nothing," Emily said. "You just look a little different today."

"So do you!" said Aaron. "Let's go to the Black Hereford. It's a lot better than it sounds."

"Okay with me," Rachael replied.

"Don't they have a McDonalds here yet?" Emily asked.

"Oh sure, but the Hereford is pretty private, and we need to hash out our plans. Unless you've changed your minds?"

"Do we have a choice?" Emily asked.

"I guess not." Aaron replied.

They spent two hours at the restaurant and came up with an elaborate plan. School was out for the year, and summer jobs were a couple of weeks off. Emily and Rachael would spend a week in Eastern Oregon learning all that Aaron wanted them to know about the forest. Then, when fall term started, they would teach Aaron everything they wanted him to know about saving the forest. Their schools were only 30 miles away from each other, and it would be easy to communicate. Once they saw the other point of view, the agreement would be over. No one would say a word about what had happened on the mountain, or about the deal they had made. Emily asked again if they couldn't forget the whole thing.

"If you're afraid to look at the other side of what you're doing, just leave." Aaron said.

"We're not afraid," said Emily. "You should be afraid because in six months, you'll be up there putting spikes in the trees with us."

"I don't think so, Emily," said Aaron. "I think that in a couple of weeks, you'll be sawing the trees down with a 30-inch chainsaw."

"So," Rachael quipped. "What time do we have to be up in the morning?"

"Be out in the parking lot at 4:30," Aaron stated.

"No, really, what time do you want us ready?" asked Rachael.

"I'm not kidding," replied Aaron. "The head forester is cruising timber on Highland Butte, which is about two hours into the mountains."

"All right," Emily answered, assuring Aaron that they would be ready to go in the morning. They left the restaurant, and Aaron dropped them off at the motel.

Emily and Rachael called home and made sure everyone knew they were spending the week. Emily's dad quizzed her about the change in plans, but was satisfied that she just needed a little time away. Rachael wasn't concerned about her parents, but was scared to death to call Herb. Herb asked her if anything had happened, and she assured him that everything was fine. She promised the pictures just as soon as they returned to Eugene. He warned her about the secrecy of FF being maintained, and she was convincing when she assured him no one would ever find out about what they had done.

Aaron asked A.J. if he could take a week and show some friends how the mill worked, all the way from cruising to shipping out the finished lumber. "Who wants to know how a sawmill works, Aaron?" asked A.J.

"Just a couple of girls who are going to write a thesis on lumber production." Aaron rarely lied, and never to his dad, but if he told his dad he had caught someone spiking trees and hadn't turned them in, it would be worse than a small white lie.

"Are they good-looking, Aaron?" asked A.J.

"No, not really. It's just a favor," said Aaron.

"Sure," A.J. said. "But be ready to bust your ass when your week's up; college isn't cheap."

"I know, Dad," Aaron promised. "I'll work extra hard for the rest of the summer." They both smiled.

Roderick Larios

Chapter IV

It was 4:30 A.M. on Monday morning, and there was no sign of Emily or Rachael. Aaron waited until 4:45, and finally the girls stumbled out of the motel. "Good morning," Aaron said with a smile. Neither Emily nor Rachael acknowledged him. Emily slid in the pickup and sat by Aaron. She was wearing the same clothes she had on when Aaron had first seen her, except this time she wore hiking boots. Rachael also wore the same clothes, hiking boots, and a Dallas Cowboys hat. They were five miles out of town when Aaron spoke. "I figured you for a Forty-Niners' fan, Rachael."

"It's not my hat, Aaron. I am a Forty-Niners' fan."

"Good guess, huh?" said Aaron.

The sun started to come up as they headed up the gravel road to Highland Butte. Aaron pulled a bag from behind the seat and pulled out a thermos of coffee and three cups. He pulled over and poured the coffee. Emily managed a smile, and Rachael had a look of relief on her face. He handed each of them a roll, which they gladly accepted. "Sorry," said Aaron. "This is the best I could do. I made us a lunch and brought a jug of water, so at least we won't starve."

"Thanks," Emily said. "The 4:30 A.M. thing made me forget all about food. Do people who work in the woods really have to get up this early?"

Aaron explained about how the woods were seldom next door to the sawmill. He told them about his first year at the mill, when he was a whistle punk. "I had to get up at 3 A.M. and ride 70 miles in the back of a crummy with a bunch of loggers; by early afternoon, I could hardly stay awake."

"What did you say you were?" Emily asked.

"A whistle punk. I sat halfway down the hill by the choker setters. When they gave an order, I'd beep the bug, and then the donkey operator knew whether to give them slack, or pull the logs up the hill, or stop, or a handful of other orders." He explained that the bug was a thing with two wooden handles with metal contacts. When it was squeezed, the horn on the donkey sounded. There was a wire running from the whistle punk to the donkey, and one had to be careful not to get it tangled. While Aaron explained that the donkey was a machine with a motor and a winch, and not some poor, overworked animal, he noticed that both girls seemed mildly interested. He went on to explain how the choker setters hooked up the logs, and how they bumped the limbs and decked the logs when they reached the landing. He started to tell them about log hauling and caught himself." I'm getting ahead of myself. You'll see all this stuff tomorrow, and it will be easier to explain."

"Do we have to get up at three?" Rachael asked, in a manner that Aaron perceived to be whining.

"No, they log all day. We had to get up early today so that we could tag along with Jed."

Jed Marshall's pickup was pulled off the road right where he had said it would be. He'd been the head forester at Douglas Timber since it was formed. He wasn't very outgoing and not too friendly, but Aaron loved the guy. He had taught Aaron a lot about the woods, and forestry in general. The first thing the girls saw was the big bumper sticker that read, THIS FAMILY SUPPORTED WITH TIMBER DOLLARS. Emily said something under her breath. Aaron ignored it.

"Morning, Aaron."

"Morning, Jed. I'd like you to meet Emily and Rachael. They want to learn everything there is to know about the forest." Emily bit her lip to keep from making some smart-ass comment. Rachael gave her that "Oh, brother" look, and she returned it.

"The first thing you need to know about the forest is that it's a renewable resource," Jed said. "But it needs to be managed properly."

"And that's your job?" asked Rachael.

"Only partly", Jed replied. "It's really everybody's job. If everybody that used the woods depended on it for their livelihood, like we do, they would understand that the people who harvest the timber are the ones who most want it preserved." Emily wondered if Aaron had said something about their views.

"Let's get moving," said Jed. "I'll explain best as I can as we go. Try to keep up."

"Mind if we use the bushes before we get started?" Emily asked. "Aaron filled us full of coffee on the ride up here."

"Go ahead, but watch out for bears," Jed joked.

"Not funny, Jed. Rachael's already had a bear encounter," explained Aaron.

As soon as the girls were out of sight, Jed spoke. "You're a lying little shit, Aaron." Aaron's heart almost stopped. How could he know what the girls really were?

"What are you talking about?" Aaron asked.

Jed replied, "You told your dad they weren't good-looking." Aaron breathed a sigh of relief.

They started up a steep slope and through a stand of lodge pole pine. Jed kept moving in a straight line by using his compass. He flagged the line with red plastic ribbons, which he hung from limbs and bushes. He explained that the first order of business was to set boundaries of the area that was to be cruised. He explained that when they did a rare clear-cut, they only worked 40 acres at a time, and marked the trees that were to be left standing. He went on to explain that they would be marking trees for a selected cut, and the area to be cruised would be an entire section, so that each border was a mile long. "How long will it take to cover the entire area?" Rachael asked.

"Not that long to cruise it for volume, but about 10 days to mark all the trees that we're going to take," Jed answered. Jed was marking the last boundary when they started, and by 8 A.M., they were ready to start marking trees. He gave Aaron and Emily each a can of white spray paint, and handed Rachael a bag of orange ribbons and a stapler. Jed and Rachael moved along the west edge of the section, and Jed pointed out the trees to be marked. Rachael stapled the ribbons to the trees, and Aaron and Emily came along behind them and painted two marks on each tree, one head-high and one knee-high. Each log hauled to the mill from this sale had to have the white paint mark on it, and each stump that was left also had to have the white mark.

Rachael couldn't resist. "Do you ever feel bad about cutting trees down?"

Jed laughed. "Do you ever feel bad about eating an ear of corn?"

"Come on, Jed", said Rachael. "It's not the same."

"Why not?" Jed retorted. "Like I told you, the first thing you need to know about the forest is that it's renewable, just like a cornfield. If we don't cut these trees, they will one day burn. Fire and chainsaws are the two ways that the forests are managed. Would you rather give people jobs and material to build their homes, or watch this forest go up in smoke?"

"Never thought of it that way," Rachael said.

By noon, they had come to the north boundary, and were overlooking a clear cut that had been harvested three years prior. There were five deer about 100 yards out in the clear cut, and at first, no one saw them. Emily saw them first and motioned to Rachael. They watched them for a few minutes; the two girls thought they were beautiful. Emily asked, "Why aren't there as many trees here, Jed?"

"Well, Emily," Jed explained. "We clear cut about 120 acres three years ago. There was some bug kill, and we needed to get rid of it. As you can see, the new trees are off to a good start, and the trees we left standing will grow a lot faster."

"So this is a clear cut?" Emily asked.

"It doesn't look near as bad as it sounds."

"It's really a sound practice," Jed offered. "But with all the media hype and all the tree hugging assholes, we try to do select logging when we can."

Aaron cringed and waited for the fireworks, but no one said a word. "Let's go eat," he said.

After they ate, they made two more passes, marking the selected trees. Toward the end of the last pass, Emily and Rachael started to lag behind a bit, so Jed and Aaron slowed their pace. They had made a lot of progress and figured they could coast the rest of the way. When they were done, Jed said good-bye and headed off to see his daughter's track meet.

Aaron, Emily, and Rachael sat on the tailgate of Aaron's pickup and polished off the rest of the water. "Well," Aaron asked. "What do think of forest harvesting so far?"

Emily answered. "I think today was a lot more fun than tomorrow will be; at least we didn't see any trees die today."

"We will tomorrow," Aaron replied. "But wait and see before you condemn the entire practice of logging. It's really neat to watch."

They got into the pickup and, as usual, Emily sat next to Aaron. Shortly after they started for home, both girls were asleep. Emily's head was resting on Aaron's shoulder, and he liked the way it felt. He tried to fight the feeling, but he knew he was falling for her. Things were a mess, but he didn't care. He was glad it had all happened; glad he had had the chance to meet her. Twenty-two, and he had never had a serious relationship. He made a vow to himself that he wouldn't until school was finished, but that was now out the window. He hoped she had some feelings for him also.

About 10 miles short of LaGrande, Emily awoke and slowly removed her head from Aaron's shoulder. She was embarrassed and quickly looked at Rachael, who was still fast asleep and leaning against the passenger side door. "I'm sorry," she apologized. "I was a little more tired than I thought. I didn't mean to use you as a pillow."

"No problem. I've been used for worse things than a pillow."

"How much further?" Emily asked. "I feel like I could sleep for a week."

"We'll be there in a few minutes," Aaron replied.

Aaron pulled his truck into the 76 station just as Rachael was waking up. "Where are we?" Rachael asked. "Did I sleep all the way to town?"

"You sure did," Aaron said. They went in and got soft drinks and snacks, and then drove to the motel. Aaron suggested they get together later for a beer, but both girls asked for a rain check.

"How about tomorrow? After we see all the trees get the ax," Emily suggested.

"That's fine," he replied. "How about picking you up at seven in the morning?"

"Sure," Emily said. "Beats 4:30."

Aaron made it to the high school just in time to see his sister run a leg of the 400-meter relay. Her team won by a good 10 yards, and Aaron was the first to congratulate her. "Nice job, kiddo. You're a lot faster than you were last year."

"Thanks, Aaron", Mary replied. "So who are the two girls?"

"Boy, word travels fast in these parts! Who told you about the girls?"

Mary thought for a moment. "Can't remember if it was Dad or Jed. They were both talking about you and some girls who were doing a report on the timber industry. Jed was telling Dad that your taste in women is extremely poor if you don't think they're good-looking."

"Their taste in women is different than mine," Aaron answered. "They're old."

"Come on, Aaron", Mary said. "They're not that old. What gives?"

"How about I tell you later. It's kind of a secret."

"So which one do you like?"

"I don't like either of them. They're just friends."

"Aaron, you're BS'ing, but I won't pry. Just let me know when you figure it out."

"Sure, Mary. If I ever figure it out."

There was a huge flash of lightning, and the thunder, which followed immediately, was deafening. Emily ran from the car to the motel room with the bag of fast food tucked under her arm like a football. She got into the room just before it started to pour. Rachael was on the phone and put her index finger over her lips. Emily mouthed, "Who is it?" but Rachael ignored her and listened intently to the person on the other end of the phone. When she hung up, Emily asked again, but Rachael just sat there and stared at the wall.

Finally she spoke. "It was James Sheldon. He called to remind me that if I were convicted on all counts of the potential things I could be charged with, I'd be subject to 100-plus years in prison." Emily went to Rachael and put her arm around her.

"That would make you 121 when you got out." Rachael didn't smile. Emily could see the fear in her eyes, and felt a lump in her own stomach. She started to make another remark to Rachael, but instead sat on the edge of the bed and held her head in her hands.

"What am I going to do, Emily?"

"Don't you mean what are *we* going to do?"

"Yes, that's what I meant." Emily pondered the question for a moment and then spoke. "I don't think we should do anything. Let's just finish the timber tour and go home. I think if the FF is going to treat us like this, I may hand in my resignation."

Rachael wanted to tell her about the file on each member that would be turned over to the FBI if they wanted to quit, but she couldn't bring herself to do it. She would tell her later, when the time was right. "I think you're right, Emily. Let's just do what we have planned and worry about James and Herb when we get home."

Aaron was waiting at the motel at a quarter to seven the next morning and saw the blind in Room 114 move. Shortly afterward, Emily and Rachael emerged, wearing the same clothes as the day before. "Good morning. You're ready early. Must be excited about today's adventure."

"Yeah, something like that," Emily muttered. This time they headed south out of town toward North Powder. About 10 miles down the freeway, Aaron took an exit and headed west toward the Wallowa-Whitman National Forest. The gravel road soon turned to dirt, and clouds of dust boiled up from behind the pickup.

It was mid-morning by the time they pulled into the log landing area. As they got out of the truck, the horn on the donkey sounded four short, loud blasts. "That means he is going to give the cable some slack, so that the choker setters can attach their chokers to the main cable. Next we'll hear one honk, and that will be a signal to pull the logs up to the landing." The horn sounded, and the three of them moved toward the edge of the landing and watched the cable pull half a dozen large logs up to the landing.

When they got to the top, Aaron explained, "These logs were cut to length by the fallers, and when they get unhooked from the choker cables, that guy over there with the chainsaw is going to bump the limbs."

"Going to what?" asked Rachael. "Cut all the limbs off the logs so they will fit on the truck," Aaron said. "Watch how fast he is." In a matter of minutes, the logs were stacked neatly in the log deck, awaiting the next truck.

"Let's start from where we left off yesterday. Jed or one of the other foresters marked these trees about a month ago. This is a private sale we bought from the Campbell Land and Cattle Company. It backs up to the national forest, and there is about a million board feet of timber that we are harvesting." He went on. "Let's go around the other edge of the landing to where they're falling."

They started down the hill to where the chainsaws combined for what Emily and Rachael thought was an eerie medley. "That's music to my ears," Aaron shouted over the sounds of the saws. The first tree they saw go over was a large Ponderosa Pine, about two and a half feet in diameter. The ground shook when it hit. The faller quickly measured the log and cut it to length. He then cut off any limbs that were easy to get to, and then headed for the next marked tree.

Aaron led them back around to where they were setting chokers. "This is really dangerous," he explained. "If one of these loose logs gets dislodged and starts to roll down the hill, those guys could be in big trouble."

"I prefer this job, myself." He pointed to a guy sitting on a stump about 100 feet up the hill and out of the main cable path.

"Is that the punk guy you told us about?" Emily asked.

" Right. He's the whistle punk. Watch when one of the setters yells at him."

"Slack," the nearest setter yelled, and the whistle punk's hand squeezed the bug four quick times. They heard the horn on the donkey echo down the hillside and saw the main cable go limp. They watched as the chokers were attached to the main. Another command, another blast on the horn, and the logs moved slowly up to the landing. They followed the logs up the hill, and were all out of breath when they reached the landing.

They sat in the truck and ate a sandwich, watching a truck getting loaded with the freshly cut logs. After the truck pulled away, the donkey operator shut the loud diesel engine off and walked toward them. "How's it going, Aaron? I thought you'd be out roaming the hills with Jed."

"Nah, I did that yesterday," Aaron replied. "Hank Barnett, this is Emily Alcott and Rachael Mayfield. They're learning about the timber industry."

"That's great," Hank said. "We need people on our side of the fence. A lot of young people are really stupid about this industry, out bitching about cutting down trees and then going home to their fucking log cabins up in the woods."

Aaron shook his head. "Thanks, Hank."

"Sorry; I can't help myself." Hank turned and walked over to the skid trail and stared off into space.

"Boy, now there's an attitude," Rachael remarked.

"He's a bitter person. He has three brothers, and they all used to work in the woods. Now it's just him. They are from Oakridge and when the government all but stopped logging in national forests, there was no log supply for the mills in that area. He's the only one to get a job in the timber industry. The state re-trained two of his brothers for the high tech industry, but neither of them could get a decent job. His oldest brother is a cook in an all-night restaurant, and is going through a divorce. Basically, the policy in the nineties tore his family apart."

"Why couldn't they find work in the woods?" Emily asked.

"There were thousands of timber jobs lost during the last administration. Whole towns have withered on the vine. The high tech industry was gonna save the day, but that industry is on its ass now, just like timber." A series of beeps sounded on the horn, and Hank fired up the motor on the donkey and continued hauling logs up the hillside.

"Is this what you would call clear cutting?" Emily asked.

"It's kind of a modified clear cut. When it's steep like this, you have to use this method to get the logs gathered. See how there are several different paths dug out and all the trees left between them? They will thin what logs they can get to, and re-plant the cut areas. They would like to clear cut, but you know the deal on that. We're doing a job on our own land that's a true selected cut. It's not steep like this is. They use skidders to get those logs to the deck. It's only about an hour from here; why don't we go take a look?"

"Fine with me," Rachael said.

"Me too," added Emily.

They headed back toward LaGrande, and then north on Highway 82. Just north of Imbler, they took a logging road to the east. About 10 miles up the road, they came to another log landing area. This one looked totally different than the other. There were no donkey or log loaders there, only neatly stacked logs in several different piles. "Take a look at that stand of trees," Aaron pointed. "What do ya think?"

Emily answered, "I think it's a shame they're getting cut down."

"That area has just been logged. That's what a selected cut looks like after it's been logged."

"Really?" Rachael asked.

"Really," Aaron answered.

The Cat was pulling a couple of big logs through the trees that were left standing, and Aaron explained. "As soon as the logs are picked up, all this brush will be piled up and left to dry, until it can be burned this winter. It'll look like a park when they're done burning and re-planting. They could, and would, do this on federal land, but thanks to the liberals, that won't happen."

"I wondered when you were gonna use the 'L' word," Emily said.

"No use calling a spade a heart, Emily. For whatever reason, liberals want to shut the woods down for good. You should know that; you're helping. Christ's sake, you're risking jail time, and you don't know anything about what you're trying to kill."

Aaron saw a tear run down Emily's cheek, and he felt horrible. "I'm sorry I get carried away like that. The whole thing just drives me nuts. I'll shut up until you've had a chance to see the whole operation." Emily turned and watched as a funny looking truck pulled up to a pile of logs.

She wiped her face with her sleeve and turned back to Aaron. "What's that thing?"

"That's a self-loader. This is really neat to watch." The truck driver lowered the stabilizers and climbed up the side of the truck and sat in a seat with a bunch of control handles in front of it. He began moving the handles, and the big jaws grabbed the first log and tossed it around like it was a toothpick. Soon, the truck was loaded, and the driver climbed back down the side, chained his load, and jumped back in the cab. The four big stabilizers retracted, and the truck roared off down the road. "Wasn't that cool?" Aaron had a big grin on his face.

Reluctantly, Emily answered, "That was cool."

"Yeah. That was pretty neat," added Rachael.

They jumped back in the truck, but instead of heading back the way they had come, Aaron drove further east.

Emily sounded a little perturbed. "Don't tell me you're cutting more trees down somewhere else."

"As a matter of fact, we're cutting trees down at a couple of other places, but that's not where we're headed. Jed always tells me that there are two ways to manage the forest. I just want to show you the other."

Rachael volunteered, "You're talking about fire, aren't you?"

"That's right. How did you know that?"

"Jed told me that when we were marking trees yesterday."

"It's not that far; it'll only take us an hour or so to get there and back."

The scenery was beautiful, with stands of aspen intermingled with the pine, fir, and larch. At one point, they passed a small lake. "Any fish in there?" Emily asked.

Aaron hesitated, and then answered. "I don't really know, but if you like to fish I know the best place on earth. It takes a little while to get there, but it's worth the trip."

"I love fishing. When we weren't traipsing through the woods, my dad had us in a boat catching dinner. Where is the best fishing on earth?"

"It's about a five mile ride from our ranch. If we get time, I'll show you."

They came over a rise in the road, and the landscape changed from beautiful to pure black. The fire had been intense, and the area that was burnt went on for a good three miles. Aaron stopped the truck, and they got out and just stared at the devastation. "Well," Aaron said, "this is what Jed calls the other way to manage the forest, and I'm sure he's right. One day there will be forest here, but not for a long time. The ground here had a lot of fuel lying around, so it burnt really hot. Nothing, not even grass, will grow here for a long time."

"I suppose you're going to tell us that a logged forest wouldn't have burned."

"I didn't say that, Emily. A logged forest would burn, but if this area had been thinned, and the fuel removed from the ground, the fire would have been easier to contain. The devastation wouldn't be anywhere close to what it is." He stayed on his soapbox. "If we were allowed to thin these national forests through selective logging, the state and federal government would save millions and millions of dollars in firefighting costs, not to mention the revenue they would get from the timber. I can't believe that the extremists get away with using things like birds and fish to help destroy one of our best resources."

Emily spoke in a sarcastic tone. "You're going to have to make up your mind, Aaron. Are we extremists or liberals?"

"You're both right now, but I think you're slipping."

"Don't bet on it."

" Okay, okay…no more sparring until the tour is over."

" Okay."

They got in the truck, and Emily hesitated and waited for Rachael to get in first, but she wouldn't move. Finally, Emily slid in the middle and sat as close to Rachael as she could get. Aaron thought to himself that he had really pissed her off, but he didn't care. He was going to show them his side, and then they would get their chance to convert him into a tree-hugger. That was the deal. They drove back through the burned-off area without a word from anyone. As they passed the lake, they could see tiny white caps being tossed around by a strong east wind. Emily finally broke the ice. "So where did you say the best place on earth to fish is?" Aaron was relieved that she was at least talking to him.

"Oh, it's about five miles up in the hills, to the west of the ranch. Sometimes I hike up there, but usually I ride. It saves a little time."

"Ride what?" Rachael asked.

"Horses. They're a lot faster than walking, and give you more time to fish." Rachael told Aaron that she wouldn't be caught dead on a horse, and he waited for the same response from Emily.

"That sounds like fun. My dad and I were always going to pack in and fish some remote lake full of huge trout, but we just never seemed to get around to it."

"I think we'll be done with the timber industry lesson by Friday. If you want, we could ride up and fish Saturday morning." Emily thought for a moment. "No, I don't want to leave Rachael sitting around with nothing to do."

"Emily, don't worry about me. We'll be tired of each other by then."

"That will work out good. We can leave for home first thing Sunday morning." Aaron thought about Emily's comment. He wondered if he'd really see her again after she left. There was no way to tell her how he felt about her. He didn't have a plan, but he would try to think of something.

They arrived at the motel just before five, and huge thunderheads were building to the south, intermittently blocking the sun. "I hate those things," Aaron pointed to the clouds. "The storms here start a lot of fires."

Emily interjected, "I figured you'd blame the liberals or the extremists for the fires."

"I know one thing. The liberals and extremists would much rather see the woods burn up than see them harvested."

"Thought we weren't going to spar until the tour was over."

"You started it, Emily."

They met at the Boar's Head at 7 P.M. Aaron was already there playing foosball on a table that came right out of the 1970's. Emily and Rachael sat at a table and watched the game. They ordered micro-brews, and Rachael paid for them with a 50-dollar bill. The waitress went off to scrape up the change as Aaron joined them. "You're pretty good at that. What's it called again?"

"Rachael, I'm sure you've seen a foosball table before."

"No, I've only seen pictures in the history books."

"Very funny. It might be a bit out of date, but it's still fun."

"We have one in our basement," Emily said. "I play it all the time."

"Sounds like a challenge, Emily," Aaron said.

"I'm not very good, but I'll probably kick your butt," said Emily.

They played five games, with Emily soundly winning all of them. "So," asked Aaron, "what do you guys think of the timber industry so far? Give me an honest answer."

There was a long pause. Emily spoke first. "Well, I will have to say that I have mixed emotions. I had my mind made up to stop the tree massacre, and I'm not so sure there is one." Aaron nearly fell out of his seat, and Rachael gave her a look that could kill.

"I think you're just getting cold feet, Emily," Rachael said. "You're afraid of getting into trouble."

"That's true," she said to Rachael." I am scared of getting into trouble, but beyond that, I think Aaron has made a pretty good case so far." Aaron kept his mouth shut and let them continue.

"I can't believe you're saying this," Rachael said. "We're in a little deep. I don't think you're considering what would happen if you jump ship."

"What can happen?"

"Let's talk about this later," pleaded Rachael.

"No, tell me what could happen if I change my mind," Emily insisted. Rachael couldn't bring herself to tell Emily about the FBI files, and didn't want to discuss it in front of Aaron, so she got up and went to the restroom.

"I didn't want that to happen. I know she's a good friend," said Aaron.

"She is a good friend. I wish neither of us was involved in this."

"It's not the end of the world. No one has to know about the spiking. I'll keep our deal."

"It's not that simple, Aaron. The spiking deal isn't the first thing we've done."

"Well, I'm glad that you're keeping an open mind about things. I'll try and do the same."

Rachael came out of the restroom and headed straight for the exit. Emily got up and joined her. "What time in the morning?" Emily asked as she was going through the door.

"I'll pick you up at seven."

Aaron sat and drank another Sharp's. He rarely drank beer, and had only been drunk a couple of times. He felt like getting blasted now. He knew that Emily was in deeper than she wanted to be, and thought she was looking for a way to get out, without being at odds with Rachael. He wanted to help her but had no idea what to do. He drove home and spent a nearly sleepless night.

For the first time he was a few minutes late. Both girls were sitting on a bench outside the motel when he pulled up. "One too many last night? Those near beers are lethal."

"Funny Emily. I like the taste of beer, just not the effect."

"Commendable," Rachael said. "What are you going to corrupt our minds with today?"

"Like I told you, today we see the logs get cut up. You'll love this."

"Emily will, but I doubt I will."

"Rachael, knock it off. I just said he had some valid points. I'm not running down and registering Republican first chance I get."

"I wondered when you were going to use the 'R' word."

"I could have used the 'C' word, Aaron."

Aaron was grinning. "Let's go see the mill."

There were four big concrete trucks sitting in line at the truck entrance, their bins turning and making a crackling noise. "I thought this was a sawmill, not a concrete plant."

"It's a sawmill, Rachael. That's why the concrete trucks are here. One thing about sawmills is that there's mud everywhere. We pour concrete all the time, trying to keep our wood clean. We also go through a lot of gravel. I'm glad you wore boots. You'll need them."

They parked and went into the office. The girl behind the desk got a big smile on her face when she saw Aaron. Her name was Nancy, and she had had a crush on Aaron forever. Emily saw the way she smiled at Aaron, and wondered if there was something going on between them. She knew that Aaron liked her, and as much as she tried to fight it, she liked him. She actually felt a hint of jealousy, and felt stupid. "Hi, Nancy. How's it going?"

"Fine, Aaron. Where you been?"

"Busy."

"I can see that."

Social Manipulation

"Nancy, this is Rachael, and this is Emily. This is Nancy." They exchanged pleasantries and went into the back office. There was a nice looking lady sitting behind the desk.

"Emily, Rachael, this is the other Nancy, also known as Mom." Nancy stood up, moved around the desk, and shook both of the girls' hands.

"Pleased to meet you. Are you enjoying your peek into the timber industry?"

"Yes. It kind of opens up your eyes," Emily said. Rachael bit her lip when she heard Emily's reply, but she didn't say a word.

"We're running some pine dimension on the day shift, but at noon we're switching back to shop." Emily and Rachael had blank looks on their faces.

Aaron explained, "Dimension lumber is like two-by-fours and two-by-sixes that are used for building. Shop lumber goes into things like doors, windows and molding plants." Emily and Rachael still didn't have a clue, but they nodded their head in agreement. Aaron grinned, "You'll learn more about that in phase five."

He gave them each a hard hat and safety glasses, and they walked from the office to the sawmill. The big Salem head rig was taking four-inch cants from a lodge pole pine log. They stood on the catwalk above the saw and watched the carriage move back and forth. The dogs on the carriage flipped the logs, and the slabs fell on a conveyer and headed for the gang saw. Log after log went through the de-barker, head rig, and gang saw, and out onto another conveyer. They then passed piece by piece through the edger, and when they came out the other side, resembled two-by-fours. They passed through the trim saw and down the belt to the drop sorter. Aaron yelled over the noise and explained what was going on. Whenever anyone passed them, they gave Aaron a big smile. Emily could tell that he was well liked. Time flew as Aaron explained in detail how the computer sized up each log, in order to get the best yield. The more Emily seemed to be into it, the more pissed off Rachael became. Before they knew it, the lunch whistle blew, and the mill wound down to a stop. Rachael looked at Emily. "Why don't you grab one of those rubber mallets and grade yourself some lumber?"

"Why don't you shut the fuck up, Rachael?"

Rachael looked as though she had just been kicked in the head. She wasn't upset about Emily cussing at her; she was upset because she knew that Emily wanted out. The worst thing that could possibly happen had just happened.

They walked over to the office as the younger Nancy was leaving for lunch. She smiled as she climbed into her Chevy truck and drove off down the road. Aaron waved as he walked into the office. "Hey, Mom, you still here?"

A voice came from the back office. "I'm always here. Sometimes I think I was born here, but I love it."

"How about lunch?"

"No, I've got to finish these bills of lading. Quicker we get the wood there, the quicker we get the money, but thanks for asking."

"We'll see you after lunch. McDonald's all right with you guys?"

"We're not guys, we're girls," Rachael snapped.

"Sorry, I meant girls."

Rachael continued, "Do you mind if just Emily and I go to lunch? I've got some stuff we need to talk about, now." Emily looked puzzled, but didn't say anything. Aaron handed Rachael the keys to his truck.

"Promise you'll come back."

Rachael responded in a more mellow tone. "I promise we'll be back for the lumber drying class right after we eat."

"See ya later," Aaron said, as he turned and walked back into the office.

"I don't know if I want to hear what you're going to tell me. You know I like him, and you're mad as hell about it." Rachael pulled the truck to the side of the road and looked at Emily.

"That's not it. For Christ's sake, I like the guy; he's friggin' Greg Brady. We have a major problem brewing and quite frankly, I'm scared shitless about it."

" Okay, now you're scaring me. What is it?" Emily started to shake.

"Emily, you can't like him. You're married to FF, like it or not."

"What the hell does that mean? I'm not married to anyone or anything."

"Yes, you are, just like me. If you ever try to quit, they will expose what you've done to the FBI." Now Rachael was shaking.

"Why would they do that? if I get in trouble for trying to torch that building, so will Herb, James, and you."

"No, they won't. They are protected with airtight alibis. They have your prints on gas cans and even pictures of you pouring gas on the side of the ranger station. They won't get touched, and we'll go straight to jail."

"Why didn't you tell me this stuff before I signed up?"

"I'm sorry, Emily. I was selfish. I knew I was stuck, and I didn't want to be stuck by myself."

"Well, that's nice. So you're telling me I'm stuck in here for life?"

"No, you're stuck in here until they say you're not stuck anymore."

"There's got to be something we can do. Maybe they're bluffing. Do you know anyone who's gotten in trouble yet?" Emily asked.

"Emily, the only members of FF I know are the same ones you know—you, Herb, James, and me. The only one of us that even knows another member is Herb, and according to him, that's just one person, the one he gets his orders from."

"Well Rachael, you're the one that's really screwed here. I may get in trouble for attempted arson and spiking a few trees, but with all the shit that you've done, you would be sent away for 100 years." Rachael began to cry, and Emily did nothing to console her. After a while, Emily spoke. "I'm going to do just as we planned, finish here and worry about FF when we get home. One thing I will not do is get in any deeper. No more missions, no more evidence."

"Emily, I'm really sorry about this. Is there any way you'd just consider forgetting about Aaron?"

"No."

Aaron was leaning against the side of the office when Emily and Rachael pulled into the lot. Emily got out of the passenger side and walked over to Aaron. "Here, we brought you a Big Mac. Hope that's all right?"

"Works for me; thanks." He ate his burger, and they went back out to the mill. He showed them the dry kilns and explained how the moisture was extracted from the wood. He could tell that both girls were pre-occupied with something, and suggested they knock off for the day. "Tomorrow, I'll show you how the lumber is surfaced and how we ship it out of here. Doesn't that sound exciting?" Both girls just stared at him, and he felt about as uncomfortable as he had ever felt. "What's going on? Did somebody die or something?"

Emily snapped out of her trance. "No. I think we're just tired. You shove a lot of info into a small amount of time." Aaron drove them back to the motel.

"How about a half day tomorrow, and then on Friday we'll go to Boise for the day?"

"What's in Boise?" asked Rachael. Before Aaron could say anything, Emily answered.

"My favorite aunt for one thing," she said. "Would we have time to go see her?"

"Sure," Aaron answered.

"What I want to show you there will only take half a day anyway."

"Thanks. I haven't seen her in months."

When Aaron got home, he went out to the big gray pole barn, grabbed a couple of lead ropes, and headed out to the pasture. He caught his horse and then Mary's and led them back to the barn, putting each one in a stall. He filled their water buckets and gave each one a scoop of grain and a leaf of alfalfa. He was checking his horse's feet when Mary walked in the barn and surprised him. "What are you doing? Why do you have Duke in here?"

"I'm gonna use him if it's all right."

"Why don't you ride your own horse? Are his feet bad?"

"Actually, Mary, I wasn't going to be the one riding Duke."

"So I'll ask you again, Aaron. Which one do you like?"

"I like both of them. Matter of fact, I like all of our horses."

"Very funny. The tall one or the short one?"

"The tall one. Her name's Emily Alcott. How'd you know there was a tall one and a short one?"

"Mom told me. Said she figured it was the tall one, but wasn't sure." Mary smiled. "You going riding tomorrow?"

"No. Early Saturday morning, we're going up to Lookout Lake and see if we can catch dinner."

"Great timing. We won't be home from districts until early Sunday."

"It's not that serious, Mary. Are you going to be mad if I miss your track meet?"

"No, the Bend schools are gonna cream us anyway. I'm sure you'll have more fun here."

"Thanks, Sis."

When Emily got back from the laundromat, Rachael was already sleeping. She wanted to talk about their options, but decided to let Rachael sleep. She had just picked up the book she was reading when the phone rang.

"How's it going? Getting homesick yet?"

"Yeah, Dad, I am a little home sick, but I'll be home Sunday afternoon. Hey, guess what? Rachael and I are going to Boise on Friday. We're going to have lunch with Lynn."

"You haven't by chance met some guy, have you?"

"God, you have a sixth sense, don't you? His name is Aaron, and I promise you'd really like him."

"Can't be worse than the last one. That kid was raised by morons."

"You didn't like him because he voted for Gore. That didn't make his parents morons. You're not a moron, are you?"

"You should have gone to Oregon State."

"This guy does."

"Drive safely. See you Sunday."

"Bye, Dad."

Aaron gave them the once-over through the mill again, as kind of a refresher course. As bored as they were, they didn't object. He gave them each a set of ear protectors and took them into the planning shed. The Stetson Ross was spitting out wide pieces of shop. The automatic stacker was placing layer after layer of the high dollar wood in the stacking bin. As soon as a unit was complete, a forklift moved it to another table, where it was banded with plastic banding. Aaron explained what was going on before they entered the planner shed, knowing it would be too loud to communicate with the high-pitched whine of the planner heads. They spent a half an hour there, and when they got outside, both girls riddled Aaron with questions. He felt like a lumber professor of sorts, but had an answer for everything they asked him.

"How'd you learn all this stuff?" Rachael inquired. "You're not that old."

"I have spent a lot of hours here over the past seven years. If you think I go into detail, you should hear my dad. He knows where every bolt, chain, switch and motor is, and he loves telling you about them. If I bore you with this stuff it's his fault."

"As much as I hate to admit it, this is pretty interesting," Rachael said.

Emily looked at Rachael and began to laugh. "That sounds funny coming out of your mouth. Having a change of heart?"

"I don't know. I thought everyone who worked in the timber industry had horns and a forked tail. Now I'm not so sure who the bad guys are."

Aaron took them out to where the rail cars were being loaded. They watched as the giant forklift placed unit after unit of lumber on the long A-frame flat car. "There goes about 40,000 bucks." Aaron was grinning. "We have to pump out four a day just to pay the bills."

"I had no idea lumber was worth so much. You must be making a killing."

"You'd think that, Emily, but it's tough to make a profit. There's a lot of competition for timber, and a lot of competition from foreign countries. My dad thinks the mills here got greedy by going direct to end-users. Wholesalers used to set the market by speculating. The prices went up and down all of the time. Now they're just down all of the time. The end users set the price, and we have to live with it and figure out how to make a profit by cutting manufacturing costs. It sounds like we're making a killing, but it's tough for mills around here to make money. My dad still sells our wood to the wholesalers. That's why we're going to Boise. Our biggest customer is there. I guess we're lucky, though; we've never lost money."

They stopped off at the Boar's Head on the way back to the motel. This time, Aaron ordered a real beer. They talked about all of the things they'd seen the past few days. "So," Aaron asked. "Do I still get to join the tree-huggers' club in the fall?"

"It's not the tree-huggers' club, and yes, you get to join. As a matter of fact, you have to join," Rachael went on. "If you do change your mind in the meantime, we'll understand."

"I won't change my mind. I'll keep an open mind about things, just like you have. Besides, I want to know why people do the things they do."

"It's a long time until September," Emily added. "Why don't we wait, and see how you feel about it then."

"Sounds like you two are having second thoughts about our deal. I really want to know if there is any valid reasoning behind what you and your friends are doing. But like you said, Emily, September is a long time off."

The drive to Boise took a little over three hours, with a stop-off for breakfast. They arrived at Boise American Trading just before 10 A.M. The building was big, and looked like it was built in the 1940's. Across the street was a lumberyard with units of lumber and other building materials stacked under huge structures with roofs, but no sides. The inside of the building was very plush, with rich wood paneling on the walls and dark green wool carpeting. Two receptionists sat behind a large oak counter with an old-fashioned-looking switchboard in front of them. They were receiving calls and transferring them to the different traders. One of the women looked up.

"May I help you?" she asked.

Aaron answered, "We're here to see Phil Greenwood. He's expecting us."

She plugged one of the antiquated prongs into the switchboard and spoke into her headset. "Phil, some people here to see you. Please have a seat. Phil will be right out." Aaron sat in one of the cushy chairs in the reception area, while Emily and Rachael looked at the old logging and sawmill pictures that were hanging on the walls.

Emily asked Rachael why a company this size would use the old phone switching equipment instead of voice mail, but before Rachael could answer, Aaron interrupted. "I asked Phil that once. He told me that they liked the personal touch, real people answering the phones and taking messages. They don't even have voice mail." Just then, Phil Greenwood entered the reception area.

"Hi, Aaron! How the heck are you?"

"Fine, Phil. Phil, this is Emily Alcott, and this is Rachael Mayfield. They're doing a paper on the timber industry and wanted to see how lumber gets sold."

"Come on up to the trading floor, and I'll show you how we attempt to sell lumber. Things have been a bit slow, but it seems to be heating up a bit."

The trading floor was a large open area with groups of desks huddled together in different areas. Phil explained that each group of desks was a different department, and that each department sold different species and types of lumber. "This is where I work," Phil said. "It's the Pine Department, best place in the company to work."

"How many salespeople do you have here?" asked Rachael.

"About 70 at last count. There are five different departments, Pine, Fir, Plywood, Import/Export, and Local Sales. We cross over on some items."

"That's a lot of people under one roof," Emily said. "How many female salespeople do you have?" Phil was caught off guard by the question, but came back with an answer.

"None right now. We used to have two ladies trading, but they moved on to greener pastures. It's hard to get decent women to lower their standards enough to work here." Phil gave Aaron a "Thanks for bringing the women-libbers to visit" look. Aaron had a big grin on his face, and Phil smiled back. One of the younger traders, who had his back turned, spun his chair around just as he was dropping the "F" bomb on whoever was on the other end of the phone.

His face turned three shades of red and he mouthed, "I'm sorry" to the girls. Emily lost it and started laughing.

Rachael was nicer and bit her lip to keep from laughing. "I guess you've seen enough here. Let's walk around, and I'll explain to you what the different departments do." Phil took them on the 10-cent tour of the trading area, and then took them outside and showed them the different items they sold to the local market. Phil explained, "Most of the lumber items we sell go out of state. We sell to every state in the union, and several foreign countries, and buy from all lumber-producing states. The local sales department sells all of the material you see here."

After the tour, Aaron, Rachael, and Emily thanked Phil and left to meet Emily's aunt. Lynn worked at a travel bureau in downtown Boise. The big poster in the window showed people in swimsuits lying in the sand with tropical drinks in their hands. "That's where I'd like to be this summer." Emily smiled. "In Tahiti, sipping Pina Coladas and watching guys run around in grass skirts."

"What's the matter, Emily? Does that sound like more fun than teaching five-year-olds to swim?" Rachael responded.

"You two have it all wrong," Aaron said. "You need to go to Tahiti in January, when there's a foot of snow on the ground."

They parked the truck, and Emily went into the travel bureau. A few minutes later, Emily and Lynn came back outside. "Lynn, this is Aaron Douglas. He's enlightening us about the merits of cutting down all the trees in the forest."

"Good," Lynn said. "Clear cut, burn, and pave, right, Aaron?"

"Not exactly. We need to leave enough trees so that we can kill all of the helpless little animals that live in the woods."

"Touché."

They walked around the corner to a restaurant that served pizza by the slice. Aaron ordered two pieces, and the girls ordered a piece each. Aaron asked Lynn, "So you're a tree-hugger, too, huh?"

"No, not really. Emily probably didn't tell you that her dad and I grew up in a family that got their income from the lumber industry. Our dad owned a hardware store-slash lumberyard in Portland. He sold out and retired in '82. When Don and I were kids, we worked summers and after school there. I have a soft spot in my heart for lumber, but I also like the forest."

"Good. So do I," Aaron said. They finished lunch and walked Lynn back to the travel bureau. Lynn hugged Emily and then Rachael and shook Aaron's hand. She said her goodbyes and went back into the travel bureau. "She nice," Aaron told Emily. "Is your dad as nice as she is?"

"Almost," Emily said.

Aaron took them by a development that Phil had told him about. There were several houses in different stages of construction. They got out and went into a house that was being framed. There were no workers on the site, so they spent a little time just looking around. "Look at all of this wood. What do think they would do if there weren't any wood to build with?" He answered his own question. "They wouldn't be able to build houses if there wasn't any wood."

"I didn't know that it took this much to build a house. There's a lot of wood here," Rachael said.

"Yeah, especially the studs. Look at how many it takes for each wall. There are a lot of sawmills that just cut studs."

Emily asked, "So how many trees did it take to get all this lumber?"

Aaron looked at her and made a feeble attempt at humor. "About five."

"You're kidding."

"Emily, I have no idea how many trees are here. I guess it depends on how big the trees were."

"I guess," said Emily.

They were on I-84 by 2 P.M. and made it back to LaGrande by a little after five. Aaron dropped them off at the motel and asked if they wanted to have a beer, or see a movie. They declined, saying that they had to get their things ready to leave on Sunday morning.

As Aaron was leaving, he asked Emily if the fishing trip was still on.

"Sure," she said. "What time shall I be out there?"

"Why don't I pick you up? That way Rachael will have wheels while we're in the mountains."

"What time?"

"I'll be here at six, if that's not too early?"

"That's fine. I'll be ready."

Aaron was in the barn at 5 A.M., saddling the horses and getting their saddlebags packed. He tied a saddle scabbard to the tack on his horse and put a model 94, 30-30 in it. He put two telescoping fishing poles and enough tackle and bait to last two days into the saddlebags. He packed a lunch consisting of PB and J sandwiches, fruit, candy bars, jerky and soda. He filled two canteens with fresh water and threw one over each saddle horn. He also packed a topographic map of the area and a compass like the one his dad had used in Vietnam. He hoped Emily might give him some pointers on land navigation. For some reason, neither Jed nor A.J. used maps. They always seemed to know where they were without looking at a piece of paper. Aaron thought it would be great if he could dazzle them with some map reading skills. He tied the horses to the hitching post and headed for his truck. As he opened the driver's side door, he heard a voice from the upstairs window.

"Aren't you gonna wish me luck?"

"Mary, you scared the crap out of me! I was coming in to wish you luck before we left. I didn't want to wake you up this early."

"I couldn't sleep very well," she admitted. "I get nervous before these big meets."

"Why don't you come down and meet the tall one before we leave? I'll be back in 20 minutes."

"I sure hope you don't call her 'the tall one' to her face."

"Haven't yet."

"I'll come down for a minute when you get back."

Emily was outside when Aaron got to the motel. It was ten 'til six, and Aaron was impressed. "You must like fishing. You're out here early."

"I don't sleep well before outings like this. I guess I get a little excited." They got into the truck, and Emily hugged the passenger side door. Aaron realized that this was the first time he'd been alone with her. He felt nervous, like it was a first date or something. She asked if they would have a chance to get a cup of coffee.

"I've got a pot going at home. If you'd like, we can take a thermos with us."

"Just a cup would be great."

"My sister will be up," Aaron said. "You'll like her; she's nicer than me."

"I'll bet she is."

Mary was sitting at the big trestle table in the middle of the country kitchen downing a cup of coffee with cream and sugar. Aaron spoke first. "Mary, this is Emily. Emily, this is my sister, Mary."

"Hi, Emily. Nice to meet you."

"Nice to meet you. Aaron tells me you're running at district today."

"Yup. We're gonna see what we can do against the big schools. We'll probably get kicked, but it'll be fun." Mary poured Emily and Aaron some coffee, then warmed up her own.

"I can't believe you're drinking coffee," Aaron said. "You're going to be real nervous by the time you get to Bend."

"I'm already nervous. Be careful not to pull the reins back too hard, Duke's got a soft mouth, and sometimes he rears up a little."

"You sure you don't mind me riding your horse today?" Emily asked Mary.

"Not at all," Mary replied. "All of our horses need to be ridden more, but it's hard to find the time."

"Thanks. I've always wanted to pack in and fish."

"You'll have a blast. Don't let Aaron get you lost. He's not the sharpest when it comes to finding his way around in the woods."

"Thanks for the warning," Emily replied. Aaron suggested they get moving. He hugged Mary and wished her luck.

Emily and Mary exchanged good-byes and shook hands. The horses were ready, and Aaron dreaded the thought of teaching Emily to ride. He started to explain about putting her left foot in the stirrup when Emily mounted Duke and gave Aaron a big grin. "You didn't tell me you knew how to ride," he said.

"You didn't ask."

Aaron climbed on his horse in true cowboy fashion, and they headed up the hill, west of the ranch.

They talked as they slowly made their way into the mountains. "So," Aaron asked. "Where did you learn how to ride?"

"My mom taught me how to ride when I was little. Her sister has a farm about 20 miles from our place; I wanted to live there when I was a kid. I was always bugging my mom and dad to take me there. My mom's sister, Jeanie, and her husband always had horses and rarely rode them, so they were glad when my sister and I showed up. I spent a lot of time riding, but we never left the farm. This is great. I appreciate it."

"I'm glad we got the chance to meet, even if it was under those circumstances."

"So am I, Aaron. Can I tell you something without making you mad, or scaring you away?"

"Sure." Aaron waited for the, "*You're a nice guy, but...*" spiel. But what she said nearly knocked Aaron out of the saddle.

"I never beat around the bush, Aaron. I think I'm falling in love with you."

"What did you say?"

"I said I think I'm falling in love with you, even though we just met. I feel that way, and I thought you ought to know. If you don't feel anything for me, I'll understand. You don't have to say you like me just to be nice and polite. I just wanted you to know how I felt."

Aaron couldn't think of anything to say, and there was a long pause. Finally, he spoke. "Emily, you may not believe this, but I had feelings for you when we first met up on the mountain, and not just the feeling that I'd like to strangle you, either. I don't know that I've ever been in love, but I'd have to say I'm falling in love with you, too."

They rode on for quite a while without saying a word, and then Emily finally spoke. "Now what? Got any bright ideas about what to do about our situation?"

"No. I'm pretty much speechless, speechless and happy."

"You probably know that I have a few issues to take care of. That tree spiking wasn't the first little terrorist act I have performed."

"What else did you do?"

"Oh, not much. Just tried to burn down a ranger station."

"Boy, you are bad! I'm glad you said you *tried* to burn down a ranger station."

"I actually tried to *not* burn it down, but I doubt it matters much." Emily went on to tell Aaron all about FF and the FBI files, and how Rachael had tried to quit. She also told him about some of the things Rachael and she had done.

After a while, Aaron spoke. "Why don't we have fun, catch some fish, have a nice trout dinner, and then figure out what our next move is?"

Emily agreed. "I like the way you termed it *our* next move."

About a mile from the lake, both horses started crow-hopping and acting up. Aaron backtracked 100 feet or so, and tied the horses to a strong branch. He pulled the 30-30 out of the scabbard, and he and Emily made their way up to the crest of the next hill. Lying in a pool of blood was a freshly killed spike elk; the back of its neck was a mass of fang marks and blood.

"Another bear?" asked Emily.

"I don't think so. Bears don't kill like that. This was a cougar, and it's a very fresh kill." Aaron felt the carcass, and it was still warm. "I've been in these woods a hundred times and have never seen a predator kill, and now I've seen two in just over a week. Pretty soon, there's only going to be bears and cougars. I hope they eat each other instead of people."

Emily started laughing. "Funny, Aaron. Bears and cougars aren't gonna start eating people."

"Oh, yeah? That's what they thought when they protected tigers in India."

"Now you're scaring me, Aaron."

"I'm kidding, Emily, but I really don't think it's a matter of *if* a bear or cougar will kill someone in this state; it's a matter of *when*. Without hounds, you just can't keep the predator population in check. Let's walk the horses around this area and go catch a fish."

"Sounds good to me."

The lake was only about five acres in size. It was set in the midst of a stand of large Ponderosa Pines, with small bushes lining the bank. It was 9 A.M. by the time Aaron had all the tackle laid out. "What do you want to try first, Emily?"

"Beats me. You're the guide."

"Might as well use my favorite, night crawlers and small bobbers. Want me to bait your hook?"

"I'm not afraid of worms. I promise I know how to fish." Emily rigged up, baited her hook, and went to edge on the lake. Her first cast traveled about 50 feet, and Aaron complimented her. Before she could respond, her bobber dipped out of sight. "I got one! I got one! Did you bring a net?"

"They're not that big. We just pull them on to the bank. Want some help?"

"No, I got it. Go ahead and cast." Aaron's line had no sooner hit the water than his bobber went out of sight.

"I thought you said they were small. This is a good-sized Rainbow. I hope I don't lose it!" Emily dragged the 14-inch trout into the shore just as Aaron's fish was nearing the bank. "Are you calling these fish small? They look like monsters to me."

"These fish are the two biggest that I've ever seen caught here. I don't think I've ever seen a fish over a foot long taken out of here." They fished for a couple of hours, releasing most of the fish they caught, and just as fast as the bite had been red hot, it turned stone cold.

"That's funny," Emily said. "They just quit biting."

Aaron answered, "They know it's time for us to eat. Nice of them to take a break."

Aaron pulled out the blanket he had stuffed into one of the saddlebags, and spread it over a flat spot on the ground. He put the sandwiches and fruit in the middle of the blanket, sat on one edge of the blanket and started eating. Emily poured water over her hands to get rid of the fish, and sat on the other side of the blanket.

"You make a pretty mean peanut butter sandwich. Are they as good as they look?" Emily asked. Aaron sat there with a mouth full of food and couldn't answer for a minute.

"You know I wouldn't have turned you in for the tree spiking," he said finally. "Why did you go along with my plan?"

"I did think you'd turn us in. I was scared, and I hated you. How did you come up with your plan so fast?"

"I was scared too. Scared I'd never see you again. It was all I could think of at the time."

"So you never really planned on looking on the other side of the fence?"

"Emily, I still want to find out why people are doing these things. I want to look at the other side."

"I don't think I'll be involved much longer. I'm not getting in any deeper, and I don't think you ought to get involved at all. These people are serious."

"I'm sure they are, but I still want to find out what makes them tick. Why don't you try and stall? Just avoid any more missions for now."

"Why are you so intent on finding out about FF? Are you some kind of secret agent or something?"

"My dad doesn't talk about it much, but my real mom was a tree-hugger. When all of the environmental stuff started, she was a part of it. She was educated at U of O in the late '60's and had what my dad describes as 'radical ideas'. I just want to know what started all this...where the roots are."

"Aaron, I think you're nuts, but I'll help you. I just don't want to get in any deeper. Okay?"

"I don't want you in any deeper. We'll figure something out. Will they ask you to do anything this summer?"

"I doubt it. I won't even be around them until fall."

"What about Rachael? Where's she gonna be this summer?"

"They helped her find a nice cushy job on campus. They've got her under their thumb, so she's screwed. Aaron, is there a way we can try and help her?"

"If it doesn't get you in any deeper, we could at least try."

"Thanks."

They finished their lunch and both lay back on the blanket and closed their eyes. Aaron's head was spinning with a million questions. What now? Do we have sex? She did say she was falling in love with me. Do I make the first move, or do I wait for her? Do I go see her on weekends? Does she really like me, or is she just screwing with me? Do I really want to know what makes those people tick? Yeah, I want to know. I really want to know.

After a few minutes, Emily asked Aaron if he'd remembered the topo map and compass. Aaron got them out of the saddlebag and spread the map out on the blanket. Emily took the compass out of its case, unfolded it, and laid it on the right edge of the map. "First thing we need to do is get the map orientated. See this little arrow with the fork off to the right? That's the declination constant, the difference between true and magnetic north. You have to line up the compass like this." Aaron was kneeling right beside Emily and paying attention like an enthused school kid. She lined the map up with the compass arrow pointing north. "We need to move the map 15 degrees. There. Now the map is lined up true north."

" That's not that hard. You're good at explaining things."

"I deal with a lot of little kids, trying to teach them to swim. You're actually easier to teach than they are."

"Gee, thanks."

"I'm serious. Little kids soak up stuff fast, but things bounce off a lot of older people. See that peak off to the south? It's right here on the map. And see those two peaks over there with the saddle in between? They're right here. If we shoot an azimuth to both of those spots, we'll find out right where we are."

" Okay, just how does that work?"

"Remember this: if the degrees are under 180, you add 180, and if they are over 180, you subtract 180. That gives us a back azimuth." Aaron shot an azimuth to both landmarks and did the math in his head.

"What now, Emily?"

"First, put the compass on the peak on the map, and draw a line down the back azimuth. Then do the same with the saddle, where the lines cross." Emily was looking at Aaron and before she could finish explaining how re-section worked, Aaron put his arm around her waist and kissed her, for what seemed like an eternity. She put her arms around his neck and just stood there, hugging him. Finally, Emily mumbled something, and Aaron asked her what she had said.

"I said, 'Great kiss'. Where'd you learn to kiss like that? That Nancy girl from the office?" Emily had a grin on her face.

Aaron's face was beet red, and he didn't know what to say. "No, I've never even held hands with Nancy. I guess I must have picked up some pointers from that *Kissing for Dummies* book." Emily and Aaron both started laughing.

"Anyway," Emily continued, "where the lines cross is your location. If you're on a road, you only need one landmark. Where the back azimuth crosses the road is your location. That's pretty much it for today's map lesson. Why don't we try Kissing 101 again? That's more fun." Aaron put both arms around her this time, and as they kissed, his hands moved over her hips, and she rubbed his back. They stood there several minutes and kissed. Finally, Aaron stepped back and looked into Emily's eyes.

"We better stop. I might never want to leave here."

They were still holding hands, and Emily answered. "Well, that would solve some problems all right, but we'd probably get real tired of fish."

Aaron packed up, took the fish off the stringer, cleaned them, and put them in a plastic zip bag. He saddled the horses while Emily dipped her feet in the water. They rode very slowly back to the ranch, talking about what a good day of fishing it had been, and how nice the weather was. Aaron had no idea what was being said. His head was plastered full of thoughts that came and went like ghosts in a horror film. He had trouble paying attention to Emily when she spoke. "You look a little distracted, Aaron. Is something the matter?"

"Oh, not much. Just don't know if I'm coming or going, that's all."

"Are you all right with this? We can call this off if you're not comfortable."

"No, I just didn't think it was possible to feel this way about someone in such a short time."

"I was thinking the same thing, but every time I think it's dumb or wrong, I think about my mom and dad. They fell in love on the first date and were married in two months. That was 31 years ago. Maybe it runs in the family."

"So you wanna get married? July is a great month for weddings."

"Funny, boy! I was just trying to tell you that things like this can and do happen, that's all."

"I know, but you don't need to convince me. I feel."

It was just after six when they got back to the ranch. Aaron took the fish in and put them in the refrigerators. When he got back to the barn, Emily was unsaddling the horses. Aaron helped her, then turned the horses out into the pasture. "That was fun, Aaron. Thanks for a great day."

"It's not over yet. We still get to eat the fish you caught. You do like fish, don't you?"

"Love fish, especially trout."

"Do you want to get a hold of Rachael, and see if she wants to come out here and eat with us?"

"Sure. I'll call her."

Rachael sounded tired when she answered the phone. "You all right, Rachael?"

"Yeah, just tired. I took a road trip today. It was neat. I went to Baker City and went through a museum that was all about pioneer times."

"I heard that was neat. You want to come out and eat with us? The fishing was great."

"I'm really tired, Emily. I think I'll just eat here at the restaurant, then get some sleep."

"I really think you need to come out here. Aaron knows all about FF and the FBI files, and maybe he can help."

"All right. I'm gonna jump in the shower. I should be there in half an hour."

Aaron put a Dan Seals CD into the stereo and started pulling stuff out of the cupboard.

"You want some help?" Emily asked.

"No, I'm fine. Just kick back. I'll be there in a minute." Aaron threw a green salad together, and then beat an egg up in some milk. He put some breadcrumbs, flour, and spices into a bowl and then went in and sat on the couch by Emily. He put his arm around her and kissed her. She put her hand on his upper thigh and rubbed him very gently.

"Maybe it wasn't such a good idea to have Rachael out," Aaron said, smiling.

"Would you care if we held off on the sex part of this for a while? I want you in the worst way, but we haven't gotten a chance to know each other yet."

Aaron thought for a moment, and then answered. "You mean wait for a while, like until nine or so?"

"No, you know what I mean. I'd just like to get to know you better first."

"I know, and I think you're right. I want to get to know you better, too." They sat and exchanged kisses until they heard Rachael driving up the road.

"Hi, guys. Are you sure I'm not interrupting anything?"

"We're sure," Emily answered. "Come on in and tell me about the museum." Rachael told Aaron and Emily about her day, and then asked how the fishing trip was. She couldn't believe the story about the spike elk, and swore she'd never go into the woods again, especially alone. Emily and Rachael talked while Aaron rolled the trout in the milk, egg, and breadcrumb mix, and put them in the frying pan with the bubbling bacon grease. He cooked three of the large trout and left the other five in the fridge. When they were a golden brown, he placed them on a platter, got the salad and dressing out, and put everything on the trestle table.

"He's a keeper, Emily. It's hard to find a good-looking guy that can cook."

"Better try them first," Aaron said. "They may look better than they taste."

Emily peeled the skin off the trout and took a bite of the steaming pink meat. "You're right, Rachael, this guy can cook. So how'd you learn to cook?"

" Self-defense when I was a kid."

They talked into the wee hours of the morning, trying to come up with a plan of attack. After going over and over all of the options, they decided the best thing they could do was to continue forward with their original plans. Emily would avoid any new missions, and Rachael, who now wanted out in the worst way, would do everything she could do to avoid getting in any deeper. Aaron would get involved, but limit his involvement to joining FF. He would be involved to a point where he would be forced to break the law, and then quit. His motives were to see the other side's point of view, and to find the origin of the environmental terrorist movement.

A.J., Nancy, and Mary made it home just after two. "Well, how'd you do, Speedy?" Aaron asked. Mary looked beat, but had a huge grin on her face. "I placed third in the 200, and we won the 400 relay."

"Wow! Going to state in two events?" Emily asked.

"No, they only take the top two finishers, but just getting there at all is way better than I expected."

"That's neat, Sis! Dad, this is Emily, and this is Rachael. This is my dad, A.J."

"So how did you like the peek into the timber industry?" AJ asked.

Rachael spoke first. "It was interesting. Kind of opened my eyes a little."

Emily continued. "It gave us all the information we needed. Aaron was a good teacher. Thanks for letting us borrow him for the week." A.J., Nancy, and Mary said goodnight, and went upstairs to bed.

"We better go, Rachael," Emily said. "We've got a long drive in the morning."

Rachael hugged Aaron and thanked him for trying to help. Aaron reassured her that they would figure a way out for her. "I'll wait in the car for you," she told Emily.

Aaron stood and looked into Emily's eyes for a long time. He finally spoke. "I really hate to say good-bye. I wish you were staying here forever. When can we see each other again?"

"Any time we want to. We can meet in the middle if you want to, or you could come to the Valley."

"How about meeting in Eugene? We can watch Mary in the state finals."

"That's perfect! I could stay with Rachael, and we could talk with Herb about you getting into FF."

"That works for me. I'll find out when it is and call you. Is it okay if I call you every day? I'm gonna miss you." Before Emily could answer, Aaron put his arms around her waist and kissed her. He moved his hands across her hips and then up under her sweater until he reached her breast. She rubbed her right hand across his back and down to his groin.

"Boy, you are gonna miss me. Maybe we should re-think the 'no sex' thing."

"No, Emily. I can wait." They kissed and touched for a few minutes, and Emily left. Aaron couldn't think of anything or anybody else. He knew he was deeply in love.

Roderick Larios

Chapter V

Professor Jonathan Finley sat in the kitchen of his grandfather's tiny house. It was one of the few private farms left on the banks of Rybinskojo Lake. It was the only time the 27-year-old professor would see his grandparents. His trip from Cambridge had taken several days, and he was exhausted, but sleep was out of the question, with such a short time to visit. The train from Moscow had arrived at Rybinsk at 5 A.M., and the trip by government automobile to Krestcy had taken little more than an hour. At 6 A.M. on that chilly September morning in 1953, Jonathon would see his grandparents for the only time in his life.

Each one of the other 137 American professors shared the same ordeal as Jonathon Finley. Not all had living grandparents. Some visited cousins, aunts and uncles, or just spent time with their parents, whom they would never see or hear from again. All 137 professors shared a common goal, which was to destroy the United States of America. They would be brought together for a one-time orientation at an undisclosed location on the outskirts of Moscow. They would then be transported to Germany, cross the Berlin Wall, and leave Europe, re-entering the U.S. from several different departure points.

In August of 1922, Vladimir Lenin commissioned 10 of the brightest minds in all of Russia. Termed social architects, they were to devise a way to destroy foreign countries without the use of force. After six months of tedious argument and compromise, a plan was devised and delivered to Lenin. The plan was perfect; it was a combination of longevity and relatively low cost. The plan would take half a century to show significant results, but the payoff would be monumental for the future of the Soviet Union. The destruction of democracy in the West was one of several goals sought by Lenin, and the two main targets were Great Britain and the United States. With Joseph Stalin's assumption of power after Lenin's death in 1924, the plan was nearly abolished. Stalin's concern was with the welfare of Russia, and only after the realization that two years of planning and implementation had already taken place, did he decide to continue the program.

In early 1923, a panel of screeners selected 200 women and 200 men. Most men were university students, and the women were last-year students in the advanced education classes of local schools. They were some of the most intelligent students in Russia, and there was an assumption that their offspring would be equally intelligent. The social architects matched the men and women, each couple compelled to engage in a forced marriage. The driving force to keep each of the participants loyal to the program and to the Russian government was their families left behind, and their love for their country. Though they were deemed volunteers, it was a mandatory assignment. Their other option was forced labor and the death of their loved ones. If, at any time during the near 30-year process, there were any deviation by any participant, his or her family would be brutally murdered.

They were all well versed in the English language, and many had studied Western culture. They were smuggled to different European countries, and entered the United States as immigrants from those countries. The mission was simple: have two offspring, and have them educated as citizens of the host countries, while teaching them about Russian culture and, foremost, about the life's mission they were pursuing. One hundred of the couples were sent to the United States, 10 couples entering as Irish immigrants.

Included were the parents of Jonathon Finley. Fifty couples were sent to Great Britain, and the remaining 50 couples were sent to other democratic countries.

The 100 couples that were sent to the U.S. were located throughout the country in random fashion. Jonathon's parents were sent to Cleveland, Ohio, where the newly named Jake Finley would work some 29 years in a steel mill. His wife, Rebecca, would spend her time teaching Jonathon and his older brother, Ned, about the mission they were on, as well as the culture of their homeland. A team of three traveling caretakers oversaw the entire process. Each posed as salesmen of varied items; these caretakers would enter the homes of the participants and grade each student on his or her progress. Upon arrival, each couple mailed postcards to five different post office boxes, and the caretakers charted the different locations and scheduled their trips, each visiting every couple in a year's time. Two couples failed to mail the postcards, and the caretakers carried pictures of mutilated bodies as gruesome reminders to those who might decide to deviate from the program.

The offspring of the couples were consumed by education, spending long hours studying. Over the years, they were always leaders in their particular schools and most all earned scholarships to leading universities. Jonathon Finley entered Harvard as a 17-year-old freshman and finished his four years at number two in his class, then number one in his class at Harvard Law School.

Of the 180 children born to the couples in the U.S., nine died before reaching school age, three died during their school years, and 31 were deemed unfit to complete the program. When a child was deemed unfit, he or she was taken by the caretakers and was never seen again. Many couples became weary of the process, but were compelled to see it through, knowing that their families in Russia would suffer, should they fail.

Each student was to study law, political science, and education. Upon the return to Russia for the final indoctrination, they would learn the method to be used against the U.S. Until 1953, there was no defined direction. This was by design, as the original architects knew great changes would occur in a time span of 30 years. A panel of Russian educators designed a program that was to be used by each individual professor. The method of delivery would be different for each one, but the basic concept was the same. Each professor would promote a new vision for students at his or her respective university. Armed with the knowledge that young people were always ready for change, they would mold the brightest minds to their way of thinking. They would encourage rebellion in the name of saving the United States from itself. During the weeklong orientation, each professor would become versed in his or her particular method of delivery. All having secured positions at leading universities, the Russian educators designed programs compatible with each university. Each area of the country had its own pace when it came to social change; the new professors would help to accelerate the change, while attempting to control the direction. The idea was to create dissension by any means available. They would mold young minds to mold more young minds and create an avalanche of adversity, spreading to every university in America. The plan was painstakingly implemented, and was perhaps the greatest attempt at psychological warfare of modern times.

The government driver knocked on the large wooden door at 9 A.M., and Jonathon Finley hugged his grandparents, stepped into the government sedan, and departed for his final destination.

Chapter VI

Aaron's first week at the mill seemed like an eternity. He thought often of the time he had spent with Emily, and looked forward to the state track meet on Saturday. He called her from Bend late Friday night, and made plans to meet her and Rachael in Eugene on Saturday morning. The first week of June brought rain to the Valley, and a light snow was falling as Aaron made his way over Willamette Pass. He chose to drive by himself so that he could stay late on Sunday, and by 7:30, he had cleared the summit on Highway 58. As he passed through the tunnel west of the summit, the sky started to clear and huge white clouds intermittently parted, allowing the sun to shine through.

Aaron took the Autzen Stadium exit, and dialed Emily's cell phone.

"Hello?"

"Hi, it's Aaron. I'm heading west right in front of the stadium. Where now?"

"There's a pizza place just a little way down on your left. I'll meet you there in five minutes. It's easier than giving directions."

"Is Rachael with you?"

"No, she's with our friend, Herb. She's going to meet us when their meeting is over."

"How'd you get out of the meeting?"

"Herb doesn't know I'm here yet."

"Good. I need to talk to you before Herb does. I've got an idea."

Aaron got out of his truck and opened the door of Emily's Explorer. As Emily got out, Aaron kissed her, and she put her arms around his neck. "I've missed you, Aaron. A week is a long time when you're in love."

"No kidding! I know the feeling. Should we be out here in plain sight? What if one of your FF people see us?"

"Gee, Aaron, I don't think there's one behind every bush. Then again, maybe there is. Follow me over to Rachael's and park your truck behind the building. I'll show you."

" Okay. I'll be right behind you," Aaron said.

The building was an old brick structure, just off campus. It was an apartment type building with a cafeteria that catered to the residents and general public. Emily showed Aaron where to park, and they made their way up the back stairs and into Rachael's apartment. The living area was decorated with cinder block and pine board shelves and tables. Aaron laughed to himself at all the wood Rachael had used in her décor.

"Has Herb ever been up here? I'll bet he gets pissed when he sees all the brutalized lumber."

Emily said, "I'm sure he's been here a time or two. I think they were an item at one time, but she won't admit it."

"When do you think she'll be back?" Aaron asked. "I need to be at Hayward Field by noon."

"She should be back by eleven. Have you eaten breakfast yet?"

"No, and I'm starving," Aaron admitted.

"Believe it or not, the food downstairs is pretty good," Emily said, so they agreed to give it a try.

Aaron had a big plate of scrambled eggs and hash browns, along with a couple of slices of bacon. Emily had toast and coffee. "So what's your idea, Aaron? Did you figure out a way to save your two favorite tree-huggers?"

"You know I'm good at plans. That's why we're here, isn't it?"

"I'm all ears," replied Emily.

"What if we get Rachael pregnant?"

"Whoa, farm boy! We haven't even had any fun yet, and you want to knock up Rachael?"

"Not me, Emily! What if we fake her pregnancy, and get a doctor to verify it?"

"What doctor is going to lie about someone being pregnant?"

"My real mom's brother is a doctor in Portland. If I level with him about our problem, I think he'll help us."

"I don't know if being with child is an excused absence in the eyes of FF, but it's worth a shot. Let's talk to Rachael about it." Aaron was polishing off the last of his eggs when Emily's phone rang.

"Hello?"

"Hi, it's me. Is he there yet?" asked Rachael.

"Yes. Are you on your way?"

"Why don't I meet you at the track meet? I think Herb suspects trouble. He was acting really strange this morning. He wanted to know every detail of our little trip."

"Okay, Rachael, settle down. Where shall we meet you?"

"I'll be standing in front of Finley Hall at 11:30."

"Okay. We'll see you there."

Rachael sat on the brick planter in front of the modern political science building.

"Hi, Rachael."

"Hi, Aaron. How was the trip over the mountains?"

"Good. How was the meeting with your boss?"

"Not good. He wanted to know everything about our stay in LaGrande. I tried to convince him that it was just for fun, but I don't think he bought it."

Emily interrupted. "Did he say anything about me?"

"He wanted to know when you were going to be here. I told him you were getting here this afternoon, and he said he'd like to see us this evening."

"Didn't you tell him we had other plans?"

"No, I didn't think that was such a good idea. I told him we have a candidate for FF, and he wants us to bring him along tonight."

"What time are we going to meet him?" asked Aaron.

"Eight, if that works for everyone."

"My parents want us to have dinner with them before they head back, if that's all right with you two. They'll probably eat around six, so we should have plenty of time."

"I don't want be a third wheel. You guys go ahead, and I'll meet you later on."

"You won't be a third wheel, Rachael. Emily and I need to talk to you. Why don't you just stick with us until we meet Herbie?"

"Please don't call him that," Emily pleaded.

"I won't."

The three of them made their way to Hayward Field just as the 1,500-meter race was starting. They found A.J. and Nancy sitting in the first row of bleachers.

"You made it." A.J. smiled.

"Yeah, Dad. I found my way all by myself."

"Nice to see you again," Nancy said to Emily with a big smile on her face.

"Nice to see you again, too", Emily answered. "When's Mary's big race?"

"The first heats in about 20 minutes. They're in the toughest heat, and three out of seven teams are really fast."

"How many do they take to the final?" asked Aaron.

"Three from each of the two qualifying heats," A.J. answered. They watched as a kid from Canby crossed the finish line to take the 1,500-meter race.

Mary ran the final leg of the 400, coming in fourth out of the seven teams. She had a grim look on her face as she approached the bleachers. "Just couldn't quite get it done," she said.

"You ran a hell of a race," A.J. said. "You made up 10 yards on your leg! You can feel good about today."

"I'd feel better if we had placed, but that's history."

Aaron, Emily, Rachael, and Nancy complimented Mary on the race. "There is one good thing about today," Mary said as she was grinning. "No more track practice."

A.J., Nancy, and Mary stayed until the meet was over, having planned to meet the others at a local restaurant at six. Emily, Rachael, and Aaron went for a drive out toward the coast. "So, Rachael, Aaron has an idea he wants to run by you."

"Is it moving to Siberia? That's about the only thing that's gonna save me."

"Tell her your idea," said Emily.

Aaron paused a moment. "You're going to become pregnant, by your new boyfriend from Eastern Oregon."

"Could you guys let me out of the truck? You're both getting a little weird," a surprised Rachael exclaimed.

"We're going to tell your buddy Herb that you were madly in love, and that you got pregnant during your vacation." Aaron continued, "I have an uncle in Portland who I think will help us fake your pregnancy. They surely wouldn't want a pregnant women running around terrorizing folks."

"That's the dumbest thing I've ever heard, but who knows?" Rachael said. "It might work."

Emily spoke. "We could bust both your legs, but you probably don't like that idea, either."

"Do you really think a doctor is going to lie?"

"I don't know, Rachael, but I'll ask him if you want me to."

"Why not? I guess anyone dumb enough to get involved in this mess would be dumb enough to get knocked up by some guy she just met. You wouldn't get jealous would you, Emily?"

"Not as long as you never get within 10 feet of each other."

Aaron dropped Rachael and Emily off at Rachael's apartment, and then drove to the Best Western by the stadium. It was almost five by the time he checked in. He hurried and took a quick shower and got dressed. At 5:30, the phone rang.

"What room are you in?" It was A.J. on the phone.

"Room 207. How about you?"

"We're in 235. Why don't you come over before we go eat?"

A.J. met Aaron at the door and stepped into the hall, speaking in a low voice. "So, looks like you're getting a little serious about Emily."

"Does it show?" Aaron asked.

"I know you. You've never been excited about a girl before, and she seems like a nice kid."

"She is nice. I hope she feels the same way about me as I feel about her."

"Just by casual observation, I'd say she does, but you never know about women."

"I heard that." Nancy opened the door, walked out in the hall, and gave Aaron a big hug.

"I concur with your dad. I think she's got the hots for you."

"Mom, please."

Aaron, Mary, Nancy and A.J. met Emily and Rachael at a quarter to six in front of T.J.'s. They were out of the upscale steak house by 7:15 and said their good-byes. Mary hugged Emily this time and whispered something in her ear. Emily gave Mary that "you gotta be shitting me" look, and they both laughed.

"What was that about?" inquired Aaron.

"Nothing," Mary replied. "Just girl talk."

"Don't be too late tomorrow," A.J. said. "You've got to meet Jed early Monday."

"I know, Dad. I'll be there." Aaron, Emily, and Rachael left in Emily's Explorer and headed toward Highway 58.

"Where are we going?" Aaron asked. Emily was silent for a moment, then spoke.

"We're going to meet Herb and James. They're going to interview you, and see if you have what it takes to be an FF member."

"What's that? A lack of common sense?"

"Don't rub it in, Aaron. Remember that you're here to help us." They drove through Pleasant Hill and Lowell, and were just a few miles from Oakridge when Emily pulled into a service station.

"Is this where Herb works?" inquired Aaron.

Emily reached over and hit Aaron on the shoulder. "Knock it off, Aaron! This is serious stuff. If you don't want to meet these guys, I'll take you back to Eugene."

"I'm sorry," he said. "I'll be good. Is this the usual meeting place?"

Rachael answered, "There is a different place every time we meet. We're supposed to meet them in the convenience store at 8:00 sharp." They waited in the Explorer until 7:55, and then went in the store and started milling around.

At 8:15, Emily walked out of the store and stood by the entrance door. Rachael and Aaron each bought a bottle of water and followed Emily out to where the Explorer was parked. Rachael spoke first. "This isn't like those two. They've never been late."

Emily answered, "I hate to sound morbid, but maybe they got in a wreck on the way up here." They waited in the parking lot until 8:45, and then headed east to Oakridge.

"Maybe he meant the station here," Aaron said.

"No," Rachael answered. "We were at the right spot. Something must have happened to them. Why don't we head back to Eugene? Maybe we'll see them on the way back."

They got to Rachael's place a little before ten, and all went up the back stairs to her apartment. Rachael picked up the phone and called Herb's number, but got a recording. Rachael looked at Emily.

"I guess we'll just have to wait until tomorrow and see what happened to them."

"Tell me, Rachael, just what did Herb say about our trip? Did you give him the pictures of the spiked trees?" Emily asked.

"Trust me, Emily. I did everything right. It's like the asshole has a sixth sense or something. He knows there's trouble."

Aaron spoke up. "You want to go have a beer somewhere?"

"Not me," Rachael answered. "I'm really tired. Pregnancy does that to a person."

"So," Aaron continued. "It sounds like you want me to talk to my uncle."

"Yes," Rachael said. "It's worth a try."

Aaron dug through his wallet and found his uncle's phone number.

"Hello?"

"Hi, Uncle Paul. This is Aaron. How you doin'?"

"Good, Aaron! How's my favorite nephew?"

"I'm doing good. I'm coming through Portland tomorrow afternoon and thought I'd stop by, if it's all right."

"Of course it's all right, I'll be home after 2 P.M."

"Great. I'll see you then."

"Oh, Aaron, I almost forgot. Your grandma would probably like to see you. She's got some stuff she thinks you and Mary should have."

"What kind of stuff? That lady has so much stuff in that house! I can only imagine what she's unloading this time."

"She loves you and likes to give you things."

"I know, and I'll graciously accept whatever it is."

"You better."

"See you tomorrow, Uncle Paul."

"Why didn't you ask him?" Emily prodded.

"I thought it would be better if I asked him in person. It's really kind of a touchy situation."

"Probably a good idea," Rachael added.

"Want to grab a beer, Emily?"

"Sure. See you later, Rachael." Emily stuck her head back in the apartment as they were leaving. "I don't know if I'll be back or not, so don't wait up."

"Good luck, sister."

They drove to a small tavern south of the campus. "This place doesn't look too lively. Are you sure it's safe?" Aaron asked.

"This is the hot spot during the school year, but it looks like business dies off a bit in the summer." They played three games of foosball, and when Aaron won the third game, he quit.

"I don't blame you," Emily laughed. "It's probably tough getting hammered by a girl."

"Yeah, I guess that's it."

"Okay, Aaron, what's the matter? You're a bit distracted."

"I was just thinking about your friends. Do you think they know you and Rachael want out?"

"I don't know. This whole thing just bugs me. I wish we could just go somewhere and get away from this mess."

"Come on, Emily. Where's your sense of adventure?"

"You want adventure? Why don't we go to your motel?" she asked.

Aaron left his half-full beer on the table and headed for the door. Emily was laughing so hard she nearly fell out of her chair.

"Wait a minute! You're gonna need me."

They got to the motel just before eleven and made sure no one was watching as they entered Aaron's room. The only light came through a small opening in the bathroom door.

Emily took off her jacket and sat on the bed while Aaron slipped out of his shoes. Aaron sat on the bed and put his arm around Emily and spoke quietly in her ear. "Who would have thought we'd end up here when we first met up on the mountain?"

"I thought I was going to jail, but this is a lot more fun."

Aaron kissed her, first on her lips, then on her neck. Emily moved her left hand along the inside of Aaron's thigh and found the buttons on his 501's. She undid two of the buttons and put her index and ring fingers through the opening and massaged him while he unfastened her bra. He was kissing her neck while his right hand moved from one breast to the other, gently rubbing each one. He was starting to run his tongue down toward her chest when the phone on the nightstand began ringing.

Emily pleaded, "Let it ring."

"I better answer it. We don't want my folks coming over to check on me." He grabbed the receiver. "Hello?"

"Hey, Aaron, you watching the news?"

"No, Dad. I was sleeping."

"Quick, turn on Channel Eight."

"Hang on." Aaron grabbed the remote off the nightstand and hit the power button. He turned on the light by the bed and clicked on Channel Eight.

"Wow, what's on fire?" he asked his dad.

"It's the ranger station up by Oakridge. The whole thing is on fire. I can only guess how that got started."

There was a long pause, and then his father said, "Aaron, you still there?"

"Yeah, I'm here."

"Well, sorry I bothered you. I just thought you'd want to see this."

"Thanks, Dad."

"Goodnight, Aaron."

" Goodnight, Dad."

Aaron sat back beside Emily and put his arm around her. She was trembling, and had started to cry.

"It's all right," Aaron assured her. "You didn't do anything wrong."

"That was the same ranger station I tried to burn down. They have pictures of me pouring gas on the side of that very building. I'm screwed, and that fucking asshole Herb set me up."

"Christ's sake, Emily, think about it. We were all together at that little store. The son of a bitch set us all up." Aaron lay back on the bed and put his hands over his eyes.

Emily sat on the edge of the bed and stared at the wall. "Sure messed up our night didn't it, Aaron?"

"Well, I'm with you, and that makes this mess better." Emily turned the light off and lay down beside Aaron. She had her head on his shoulder as they discussed their newest problem. They eventually fell asleep.

The phone rang at 7:30 the next morning, and Aaron answered it.

"Hello, Dad."

"Aaron, it's Rachael. Have you seen the newspapers yet?"

"No, but we know about the ranger station. We saw it on the news last night while it was burning."

"Are you two coming over here?" she asked. Aaron turned to Emily, who was barely awake.

"Rachael wants to know if we're going over to her apartment." Emily took the phone from Aaron.

"Can you believe what Herb and James did to us?"

"I can believe it. I just never dreamed it would happen. are you coming over here?"

"Sure, we'll be over in a little bit." Emily put the receiver back on the hook and turned to Aaron. "Good morning, I had the worst dream last night."

"Me too. I dreamed we almost had sex, and then the phone rang." They smiled at each other and hugged.

Emily spoke first. "It will happen. I just hope it's not a conjugal visit." They took turns in the shower and left for Rachael's.

Rachael was sitting on the bottom row of steps in the back of her building. Emily opened the door of Aaron's truck and Rachael slid in beside her. Rachael was shaking when she spoke. "I'd like to kill Herb and James. I can't believe they did that."

"We feel the same way," Emily responded. "We're on our way to do just that."

"Great. Might as well add 'Murder One' to my list of felonies."

"Don't worry, Rachael", Aaron said. "We're just keeping murder as an option."

They drove to the small rental house in North Eugene that Herb lived in, and parked around the block. Herb's old Buick was in the driveway. Emily knocked on the door and waited for Herb to answer it.

Herb was wearing a white tee shirt, gray sweat pants, and no socks. "Thought I might get a visit from you guys," he said. "Come in."

Emily's voice was a little shaky when she spoke. "I can't believe you set us up like that."

Herb looked directly into Emily's eyes when he answered. "I didn't set you up, Emily. I didn't know anything about the fire until James called this morning."

"Oh, bullshit," Rachael yelled. "You were supposed to meet us. Where the hell were you?"

"By the way, I'm Herb," Herb said, ignoring Rachael. He stuck out his hand, and Aaron reluctantly shook it.

"I'm Aaron Douglas, and I can't say I'm pleased to meet you."

Herb went on. "I'm only going to tell you guys this once. It's the truth. You can believe me, or you can think I'm lying, your choice. Yesterday afternoon around 4 P.M., I got a call from my FF contact, asking me to stop by a meeting point south of town at seven. He knew about our meeting, and I thought he was going to give me some information he had about Aaron. I thought something was a little strange because he told me to bring James along. When we got to the meeting site, my contact was nowhere to be found. At about 7:10, two guys pulled up in what I swear looked like a government vehicle. They got out and came over to where we were sitting and asked us to get out of the car. At first, I thought we were getting arrested or something. They suggested we get into the back of their car, and we obliged. They told us not to talk, and drove us south on I-5, taking the Oakridge exit. I thought it had something to do with our meeting, but I was afraid to say anything. Anyway, we got as far as Pleasant Hill, and they pulled into a service station, and the guy who wasn't driving turned around and gave me a twenty-dollar bill."

Emily couldn't help herself. "Herb, you're lying! That's the biggest crock of shit I've ever heard."

"I told you, you could choose to believe me, or to not believe me. Do you want to hear the rest of the story or not?"

Aaron interjected. "Let him finish. If it's not true, it's still a hell of a story."

Herb continued. "The driver turned around and told James and me to go into the station and each buy a two and a half gallon gas can. I couldn't help myself, and I asked what the cans were for. I've made mistakes before, lots of them, but this was a beauty. The man driving pulled a pistol out of a shoulder holster and pointed it at my head. He said, very casually, because he told me to. That was enough explanation for me. We did as he requested. He had us each get our cans full of gas and get back into the car. They started driving back to Eugene. There we were with cans of reeking gasoline. It was about then that James pissed his pants. I don't think he was that scared; he was just afraid to ask if we could make a piss stop. They parked in the Denny's lot on I-5, and took the gas from us. The guy who wasn't driving put on a pair of gloves and put the cans on the floorboard in the front seat. They told us to keep our mouths shut about the whole incident."

"So your contact set you up, and you set us up without knowing it?"

"Emily, it's the truth. James is on his way over here as we speak. Ask him what happened. His story will match mine perfectly. There is no way we could make this up and remember all the details. It happened."

Aaron spoke up. "Did your contact tell you where you were supposed to meet us?"

"As a matter of fact, yes. He told me where and when."

"Have you talked to your contact yet?" Emily asked.

"I've been calling him since midnight, but there's no answer, not even the usual recording. I've never known where the dickhead lives, or I'd be on his doorstep right now, giving him hell."

James didn't bother knocking as he opened the door to Herb's house. "Cool! A party."

"Yeah, it's a party, all right." Emily's voice was as cold as ice. "It's a 'let's all go to jail' party."

"Sorry, Emily. I know it's not funny. Did Herb tell you what happened?"

Emily asked James to step outside for a minute, which he did. She then quizzed Herb about the appearance of the two men. What were they wearing? What color was their hair? Did either have a mustache? When she was done, she asked James to come back in the room. She than asked James the same questions. With only minor discrepancies, James described the two men the same way Herb had.

"If you're telling the truth," Rachael went on, "who the hell is setting us up, and why?"

"I don't know," answered Herb. "I thought I was gonna change the world with a handful of 30 penny nails. When I started doing this shit five years ago, it seemed like the right thing to do. I hated seeing log truck after log truck roaring down the highway. I never stopped to think about where I'd end up. I sure as hell don't want to get bent over in some shower at the state pen, but it doesn't look good right now."

They discussed their different problems, holding nothing back. Rachael was pleased that she wouldn't have to fake a pregnancy. Aaron was relieved that he wouldn't have to ask his uncle to falsify a pregnancy exam. Those were about the only high points of their conversation.

Herb and James vowed to find out who and why they were set up. Rachael, Emily, and Aaron promised to give them a chance, and not interfere. They all wanted answers, but didn't really know where to look. Herb and James knew they weren't going to perform any more acts of terrorism. Emily asked for the pictures of her dousing the ranger station with gas. She wasn't surprised when Herb told her that his contact, Ernie Roper, had all the pictures and other hard evidence, so Ernie was the big question mark. James knew that Ernie was the go-between from the main headquarters to several terrorist cells in the area. On several occasions, Ernie had bragged about being a part of what he had termed "middle management." Herb had asked on more than one occasion who some of the higher-ups were, but Ernie wouldn't give him any names. He did tell Herb that he would be shocked at some of the people in charge.

The five of them decided to take a "wait and see" approach, in lieu of a better plan. They would share information and communicate over the Internet and would meet in person on the first Saturday of every month at different locations. The first of these meetings would take place on July 7. Rachael told Herb that she was moving home for the summer, and he agreed that it was probably for the best. He and James would try to find Ernie and get some answers.

Emily and Aaron said goodbye to everyone else and went back to the motel together.

It was almost eleven by the time Emily and Aaron had picked up Emily's things at Rachael's and made it back to the motel. Aaron's parents were long gone.

"Want to go in and mess around a while?" Emily asked.

She had a smile on her face, but Aaron could tell it was forced. "It's checkout time, and we both have long drives. We better get rolling."

"All right. Why don't you follow me home and meet the family? I know you're excited about that."

"Actually, Emily, I'm looking forward to meeting your parents. After we do that, I'll take you by my grandma's house. She's a neat lady."

"Are you still going to stop by your uncle's house?"

"I think I better, since he's expecting me."

"This is some serious stuff, isn't it? I mean meeting each other's relatives and all."

"About as serious as I've ever been, Emily. I love you."

"I love you, too."

Social Manipulation

The trip to Tigard took about 90 minutes, with one pit stop. Emily's house was on the outskirts of town and set on about an acre of ground. The yard and landscaping were immaculate, and the SUV and pickup truck in the driveway looked like they had just rolled off the showroom floor. "Boy, your dad likes to keep things ship-shape, doesn't he?"

"It's not my dad, it's my mom. She loves the yard work, but she makes my dad do the cars. Come on in. Hello? Anybody here?" she yelled as they approached the house.

"Out back, kiddo. We're cooking you guys some dinner."

"Hope you're hungry," Emily said. "My dad thinks he's the king of barbecue. He's got some fancy thing that burns wood pellets; you'll probably like that."

"I'll tell you what I think after I sample the product." Emily took Aaron through the house and out onto the patio. The backyard was even nicer than the front. There were large flowerbeds and a beautiful in-ground pool.

"Dad, Mom, this is Aaron Douglas. Aaron, this is Don, and this is Trish Alcott."

"Nice to meet you, Aaron, Emily tells me your dad owns a sawmill," Don said.

"Yes sir, the best mill in Eastern Oregon."

"My dad was in the lumber business until '82." Don explained. "We sold a lot of pine boards that were made in your neck of the woods."

Trish interrupted. "Why don't you just tell Aaron some war stories? That's more interesting than lumber."

"I'll bet Aaron likes talking about lumber."

"I do like talking about lumber, but sometimes I get real tired of handling it. What unit were you with in Vietnam?"

"First Cavalry, but I don't really tell that many war stories."

"My dad was with the First Infantry Division; he doesn't talk about it much either." They talked for an hour or so and then had dinner.

It was about four when Aaron and Emily left for Aaron's uncles. "Thanks for dinner. Those steaks were about the best I've ever tasted."

"Thanks, Aaron." Don had a big smile on his face. "I can't take all the credit. Those Traeger BBQ's are the best things since sheetrock nails." Aaron said goodbye to Don and Trish as he and Emily walked out to his pickup.

"Your parents are really nice," he said.

"They put their best foot forward today. I had them properly briefed on how to act."

"Right, Emily. I'm sure they do exactly as you say."

" Okay. What you see is what you get. They are nice. Do you want me to follow you to your uncle's and your grandma's, so you won't have to bring me back?"

"No, I'll bring you back. That way, we'll have more time together."

Paul Simpson's house was in an upscale community on top of a small mountain just south of Portland. The houses were similar, all large, with beautifully landscaped yards. A Jet-black 911 Porsche sat in the driveway.

"Nice ride," Emily said. "Your uncle isn't hurting for cash, is he?"

"I've never seen his checkbook, but I think he does all right. That car is my cousin, Robby's. And I'll bet he's buying it himself. My uncle's not that tight, but he doesn't give his kids a lot of stuff."

"Is your cousin a doctor too?"

"No, but he's in his last year of medical school. The guy is the biggest entrepreneur I know. He made a thousand bucks the year he turned eight. He and some friends bought pop bottles from people at half the cash return value, and doubled their money. He was buying Microsoft stock before it split the first time."

"Do I get to meet this guy?"

"Yes, just don't get too friendly."

"Jealousy? I like that."

Aaron rang the doorbell a couple of times, but no one came to the door. Finally, Robby stuck his head around the corner of the house. "We're out back. You guys hungry?"

"We just ate. Robby, this is Emily. Emily, this is my cousin, Robby."

"Hi, Emily. Nice to meet you."

"Nice to meet you, Robby." Robby took them to the backyard and introduced Emily to his parents and his younger brother. Paul Simpson was a nice-looking guy in his early fifties. Aaron's Aunt Sara was very sophisticated. Robby's brother, Frank, looked like he could use a bong hit.

"Hey, Dude." Frank put his arm around Aaron. "How's the tree butcher doing?"

"Killing all we can, Frankie. How have you been?"

"Good. College is a little tougher than I thought it would be, but at least I made it through the first year."

"I told you to stay away from Stanford. You're a glutton for punishment."

Paul took Aaron aside while Emily visited with the rest of the family. "Have you got something on your mind, or is this just a social visit?"

"Just a social visit, Uncle Paul, why?"

"Just wondered. You don't drop by that often, so it's good to see you."

"So what's Grandma got that she's trying to unload?"

"I'm not sure, but I know she's home today."

"That's our next stop."

"How serious are you and Emily? She seems like a good kid."

"I'm in love with her. I guess that's serious."

"Like getting married serious?"

"I don't know where we're headed for sure, but I'll keep you posted. We haven't known each other very long."

Aaron and Emily said goodbye and drove to Aaron's grandma's house.

Aaron's grandma lived in an older neighborhood in Southeast Portland. Her house was modest-looking and kept up fairly well. There were miniature picket fences lining the flowerbeds, and several trees of different species spaced throughout the front yard. Aaron didn't bother knocking. He opened the front door and yelled, "Grandma, it's Aaron. Are you home?"

"Who is it?" came a voice from somewhere in the house. "I don't remember anyone named Aaron."

"Yes, you do. I'm your favorite grandson." Aaron's grandma came down the stairs carrying a vacuum cleaner, which Aaron quickly took off her hands. "Grandma, this is Emily. Emily, this is Martha Simpson."

Martha put her arm around Emily's shoulder and spoke to her in a low voice. "Any friend of Aaron's is a friend of mine. It's nice to meet you, Emily."

"Nice to meet you, too." They sat in the front room and talked for an hour. Emily listened as Martha told stories about Aaron and Mary. Aaron finally got up to leave and hugged his grandma.

They were heading through the door when Martha remembered something. "Wait just a minute. You didn't think you were leaving empty-handed, did you?"

"Grandma, you don't have to give me anything."

"Come upstairs and help me. The boxes are getting heavier the older I get." Emily followed Martha and Aaron upstairs and into what had been Katie Simpson's bedroom, Aaron's real mom. There were still U of O pennants hanging on the wall, and the room still looked lived in. In one corner of the room there were two large boxes with Aaron and Mary's names written across the top with a magic marker. "These are some of your mom's things from high school and college. You might want to go through them. I've never quite been up to it."

"I don't know that I'll ever be up to it, either. But I know Mary will." Aaron taped the tops and carried one box at a time out to his truck. Aaron met his grandmother on the front porch and gave her a big hug. Emily said goodbye, and Martha made her promise to visit again.

The sun was starting to set by the time they made it back to Emily's. The shadows had climbed halfway up Mount Hood, and the top, still in the sun, was beautiful. Emily was standing with her arms wrapped around Aaron, whispering in his ear. "I wish you could stay here; I'll miss you."

"I'll miss you, too, but I need the job, just like you do."

"Where are we going to live when we're finished with school? Providing it isn't prison."

"Emily, please stop the prison talk. We can live anywhere we want, but my vote is East."

"So is mine."

Aaron kissed her goodbye and headed for LaGrande.

Chapter VII

Aaron made it to the ranch just after 1 A.M. and parked his truck in the driveway. He was halfway to the front door when he remembered the boxes. One at a time, he carried them into the house and set them in the den. He was tired but something inside him made him take out his pocketknife and slit the tape open. There was file after file of carefully labeled folders. The first one Aaron looked at was marked "Things that need doing." The first thing on the list was "Stop the War." Aaron looked at the list in amazement. Number two was "Women's Rights," followed by "Save the Whales," "Stop World Hunger," "Socialize Medicine," "Legalize Pot," "Tax the Rich," and coming in at number eight was "Save the Trees." He put the papers back in the folder and put it back in the box.

As Aaron passed by Mary's door on the way to his room, she whispered, "Made it home okay?"

"Yeah, all in one piece. Hey Mary, come home as early as you can after work tomorrow. I've got something to show you."

"What?"

"Tomorrow, Mary. I'm tired."

"Goodnight, Aaron."

" Goodnight, Mary."

Aaron made it home from work just before five, and Mary was waiting on the front porch. "What's so important?"

"Follow me." Aaron showed her the boxes, and they both started looking through the different files.

They were silent for quite a while when Mary spoke. "Look at this. Oh, my God, this is amazing."

"What?"

"It's her journal, all the way from her freshman through her senior years. I almost feel guilty reading this."

"I feel the same way. If you find anything interesting, let me know."

They looked through the files most of the evening, stopping only for a quick bite. It was almost ten when Mary, who had been reading the journal, spoke. "Aaron, you've got to read this."

"Read it out loud."

Mary started reading. "Friday, October 17, 1969. I feel like I'm part of some bizarre experiment. Today while I was helping Professor Finley organize the weekly assignments, Professor Amy Grayson visited him. I'd met her before, and knew that she was a law professor at UC Berkeley. As in the past, Professor Finley asked me if I'd please excuse them. I left as usual, but forgot my coat in the closet by the classroom entrance. I went back to get my coat and heard the door to the main classroom close. I almost yelled at Professor Finley, but as I approached the closet door I saw Professor Finley and Professor Grayson embracing. I didn't want to embarrass them, so I ducked behind some storage boxes. They started kissing, and I couldn't help myself. I was peering over the boxes, watching them."

Aaron interrupted. "This is getting good."

"You haven't heard anything yet." Mary continued. "Professor Finley held Professor Grayson against the wall, and his hands were moving over her breast and her butt. I couldn't believe what I was seeing. They went on like that for a short while, until finally Professor Finley stopped. He told her they would continue later that evening. As strange and uncomfortable as I felt at that time, what came next was the most bizarre conversation I'd ever heard. Professor Finley asked her how the project was going. Her first comment was that the Vietnam War was a blessing, and he agreed. He went on to tell her that he had educated enough liberal lawyers and teachers to contaminate the entire legal and education system for the state. He told her that in a matter of a few short years, criminals in Oregon wouldn't do much jail time for most crimes, with all of his judges in place. He told her that it was incredibly easy to sell liberalism to young minds. She told him that they, whoever they were, were concentrating on the entertainment industry and the media. She told him that the message they were trying to send was that big business was the enemy of the common person. That it was the right of every poor American to receive government assistance. She was explicit in saying that they had the black community right where they wanted them. She said that the welfare system would keep her people voting for her boys. They talked for a few minutes about someone called the "caretaker." She voiced some concern about being caught with Professor Finley. The last thing she said before they left the room was that soon they would have the East and the West Coast under control, and just work their way to the middle.

It was the most bizarre thing I have ever heard. I thought about telling Paul what I'd overheard, but decided not to. I'd like to ask Professor Finley who he really is, but most likely I never will."

Aaron and Mary sat in silence for some time. Aaron stood and went to the window in the kitchen and stared out at the full moon. He finally walked back into the den. "What in the hell do you make of that? Did our mom really hear that stuff, or did she have one hell of an imagination?"

"Aaron, I think she heard it. Listen to the first sentence on the next page, dated October 18," Mary said. She took a deep breath and read aloud. "I didn't sleep last night. I'm afraid to tell anyone what I've heard for fear of being labeled insane." Mary looked up. "That must have been a nightmare, not being able to tell someone. Aaron, who do you think they were?"

"I don't know, but that name, Finley, sounds familiar. Its just 10:30. I'm gonna call Emily."

Aaron called Emily and told her about what they had found in the journal and asked if she knew of a Professor Finley. Emily didn't know him personally, but had heard of him. She told Aaron that she would call Rachael and find out who this guy was, and would call him back shortly. While they waited for Emily's call, Mary and Aaron kept going through the files.

Just before eleven, A.J. and Nancy walked through the front door. "Well isn't this nice? Our kids enjoying some quiet time together."

Mary Smiled. "Hi, Dad. We were looking at some of our mom's stuff. Grandma gave us these boxes full of her schoolwork. It's pretty interesting."

"Oh, yeah? What's so interesting about that stuff?" A.J. asked.

Nancy excused herself and went upstairs.

"Dad, you have to read this and tell us what you make of it." Aaron handed A.J. the journal, opened to the page Mary had read out loud. A.J. read the page, and then read it again. He looked stunned.

Finally, he spoke. "I can't believe what I just read. Who were these people?"

"That's what we're trying to find out. I called Emily, and she's calling Rachael as we speak. Rachael will probably know; she's kind of a U of O history buff."

The phone rang, and Aaron answered. "What did you find out?"

"Hi, Aaron, this is Rachael. Why do you want to know about Finley?"

"Just curious. Do you know who he was?"

"I know who he is. I had a class from him my freshman year, the year he retired."

"He must be older than the hills."

"Maybe, but he was still full of piss and vinegar when I had him. He is known as the Father of Reason at U of O. That building I met you guys in front of on Saturday morning was named after him."

"Is he the one that signed you up for FF?"

"No, but he's adamant about the environment. He changed a lot of lives, including mine. He didn't sign me up, but if weren't for him, I wouldn't have gotten involved. He preached change at all cost. Why do you want to know about him?"

"Like I said, just curious."

"If you want to know more about him, you should e-mail Herb and quiz him. He knows a lot more about the professor than I do."

"Maybe I'll do that. Thanks for the info."

"Hear anything about the fire?"

"No, have you?"

"Not yet."

"Goodbye, Rachael, and thanks."

"Goodbye, Aaron."

"What's FF?" A.J.'s question caught Aaron off-guard.

"What's what?"

"You asked Rachael who signed her up for FF. What is FF?" Aaron didn't want to lie to his dad again, but the thought of telling him what Rachael and Emily had been doing was too much.

"It's a group called Forgotten Friends. It has something to do with helping people who have drug or alcohol problems."

"I've never heard of that one. Seems like there's a group for just about everything." A.J. went upstairs, and Aaron and Mary sat in the den until midnight, going through the files. Mary went upstairs first, followed shortly by Aaron. Neither Mary nor Aaron slept well. The same question went through their heads. "Who are these people?"

Aaron was late meeting Jed the next morning, and got an earful. Most of the ass chewing went right over his head. With every step, the same nagging question went through his head. Jed sat at the edge of a clearing, waiting for Aaron to catch up. When Aaron made it to the clearing, Jed got up and walked over to him.

"What the hell's a matter with you?"

"Nothing. Why?"

"Don't bullshit a bull-shitter, Aaron. Now what's bothering you?" Aaron sat on a log and told Jed about the journal and the bizarre story of the two professors. He asked Jed what he thought.

"That's the damndest thing I've ever heard. You're not feeding me a line, are you?"

"No, Jed, I swear. Ask my dad; he read it, too." Jed sat for a while, digesting the story.

Finally, he spoke. "Must be some kind of a government experiment or something."

"What about some other government's experiment?"

"I wouldn't go around telling everybody about this. You might get a visit you don't want."

"Not that many people know about it. I'd only tell people that I trust."

Aaron made it through the week, and by Friday, the question about the journal had slipped behind his thoughts of Emily. He called her every night and had made plans for a trip to Portland the following week. When he talked to Emily on Sunday afternoon, she asked him if he'd had any contact with Herb. He told her that he hadn't, but would email him, and see if there was anything new.

Emily told Aaron about a story in the Sunday paper about a body that was found in the river just north of Eugene.

"Do you think there's some connection?" he asked her.

"I don't know, but if there is, Herb will probably know."

"Emily, I'll email him as soon as we hang up."

"Herb, anything new there? Please mail back any new info you might have. What do you know about a professor by the name of Jonathon Finley? Thanks, Aaron." Aaron waited for a response but nothing came back. He shut the computer down and opened up one of the boxes of files. He took the journal out and read the part about the two professors again.

He read a part about his mom and dad meeting. It was during a war protest in Portland. A.J. had just gotten back from Vietnam and was working for a wholesale lumber company. The office was located a block away from the park where the protest was taking place. He had to laugh when he read the part about his mom seeing his dad for the first time. "A guy asked me why I was wasting my time carrying a stupid sign around all day. I told him I wanted to make a difference. He told me that all the kids that were dying in Vietnam wanted to make a difference too. Then he asked me if I'd have a beer with him later. I don't why I said yes, but I'm glad I did. We had a wonderful time. I like this guy. He said he would call me. I hope he does."

Aaron put the journal back into the box and fired up the computer. Still nothing from Herb, so he shut the computer down. He thought of his trip to Portland scheduled for the following week, and wondered if "it" would happen this time. Emily had mentioned that her dad wanted to take them fishing. As much as he loved to fish, Aaron wanted to just spend time with Emily.

The following week was uneventful, until Wednesday evening when Aaron turned on the computer and saw the email from leftwing80550. Herb's first sentence set the tone. "Aaron, sit down. It was Ernie Roper, the body from the newspaper story. Whoever burned down the ranger station must have killed Ernie. I don't know if any of us are linked to a murder, but this doesn't look good. Professor Finley is my mentor. He is the brightest person I've ever met. Why do you ask? If you have any info, please email back. See you in early July. Herb."

Aaron walked outside and hiked about a mile into the woods. He sat on a big pine log and stared off into space. He had options, but didn't have any idea which one might be the right one. He thought about calling the state police or the FBI, but didn't want to implicate Emily. He would just wait and see what happened next. He walked through the woods and contemplated the situation. He asked himself if he would have been better off if he had never met Emily and Rachael. The answer was the same as always; no matter what, meeting Emily was the best thing that had ever happened to him.

Aaron worked in the mill on Friday, and as soon as the four o'clock whistle blew, he was headed for Portland. The usual five-hour drive took just under four and a half hours. Just before nine, he pulled into the Alcott's driveway. Emily met him at the front door and wrapped her arms around him. He kissed her, and just stood there holding her for a few minutes.

"I'm glad you came," she said. "Have you eaten yet?"

"I grabbed a burger in The Dalles, but I'm still a little hungry." They went into the kitchen, and Emily made him a sandwich.

Don and Trish walked into the house just before ten. "Hi, Aaron. Nice to see you again."

Don shook Aaron's hand. Trish said "hello," and tossed Emily an approving look. "So, Aaron, are you up for a little salmon fishing and crabbing in the morning?" asked Don.

"I've never been salmon fishing or crabbing, but it sounds like fun."

"Good", Don said. "We'll leave at 5:30, if that's not too early." Don and Trish went into their bedroom and Aaron finished his sandwich. He and Emily walked out into the backyard and sat in a lawn swing that faced the pool.

Aaron put his arm around her. "Did you hear who the floater was?"

"Yeah, Rachael talked to James. He's scared to death."

"Why do you think they killed him?"

"The paper didn't say he was murdered. It just said he drowned."

"Come on, Emily. You know somebody killed him, and then threw his body in the river."

"That may be, but you don't know that for sure. Maybe he got drunk and fell into the river. That has happened."

"I hope you're right." Aaron kissed her and slid his hand under her tee shirt and rubbed her back.

She kissed him on the neck and whispered into his ear. "My mom made you a bed on the couch. I hope that's okay?"

"That's fine. I'd feel weird if we did anything here anyway."

"I wouldn't, but I know what you mean. We better get some sleep." Emily showed Aaron where everything was, and then went to her room.

Aaron was already dressed when Emily came in to wake him the next morning.

"Beat me to the punch, huh?" She smiled at him.

" Geez, Emily, the day is half over. You know what time I have to get up for work."

"You're warped." The trip to the mouth of the Columbia took about two hours. Aaron helped Don launch the big jet sled. They headed up river, and dumped the crab traps into the murky water.

Don took the boat out near the bar, and the swells were a good five or six feet high. They rigged their poles and started trolling in amongst the other boats. Emily was the first to get a fish on. She fought it for several minutes and finally got it close enough for Don to net. It was about a 15-pound Silver. Before Aaron could congratulate her, the tip of his pole bent toward the water.

He grabbed the pole from its holder and gave a big tug. The fish was unlike any he'd ever caught. "Wow, this thing is huge!"

"Keep your tip up," Emily yelled. Aaron fought the fish for 20 minutes before they even got a glimpse of it.

"Hang on, Aaron," Don yelled. "That's a Chinook, and a good one." Finally it was close enough, and Emily netted it. It was the biggest fish Aaron had ever seen.

"That was fun! I had no idea these fish were this big."

"Most of them aren't," Don said. "I'll bet this fish goes 40 pounds." They fished until noon and had four fish in the boat. As Don turned the boat up river, Aaron caught a big whiff of exhaust and was quick enough to get his head over the side of the boat.

"Welcome to ocean fishing," Emily was laughing. Aaron turned to her with a big grin on his face.

"That wasn't much fun, but it's worth it." They pulled the crab traps up one by one, and sorted the keepers out. They had four over their limit and threw them back. They were loaded and ready to roll by 1 P.M.

"Thanks a lot, Mr. Alcott. That was fun."

"You're welcome, but call me Don."

" Okay. Thanks, Don."

Don brought a big pot from the garage and filled it with water. He added some salt and spices and put it on a propane burner. When the water started to boil, they threw a dozen of the crabs into the pot. Don took the fish into the utility room, and he and Emily cut the fish into fillets and steaks. They saved two nice fillets and packaged the rest. "We'll freeze your fish, and you can take it home in a Styrofoam cooler."

"Thanks, Don, but I don't need all that fish. Just a couple of pieces would be great."

"If you don't want all of it, give some away. It's a great way to make friends."

They cooked the crabs in three sessions. When they were done cooling down, they were put into a large tub of ice. Don couldn't resist. He grabbed a crab and pulled one of its legs off and broke it open. Emily, Aaron and Trish followed suit.

They were finishing off their initial tidbit when a newer Suburban pulled into the driveway. A good-looking girl got out of the passenger side and made her way to the front of the garage where the crabs had been cooking.

"Hi, Dad; Hi, Mom; Hi, Emily; and it's nice to meet you, Aaron."

Aaron couldn't believe how much Sara looked like Emily. "Emily told me you two looked a lot alike, but I didn't know you were twins."

"They're not," said the tall handsome guy who was driving the Suburban. "Emily's a lot meaner than Sara. You must be Aaron."

"And you must be Kevin. Emily's told me a lot about you guys." They shook hands and talked about how good the fishing and crabbing had been. They took the party to the backyard, and Don put the two fillets on the Traeger and turned it on to smoke.

"Where's Mikey?" Emily asked. Kevin answered her with a big grin. "Sleeping in his car seat, and please leave him there until he wakes up. That kid has way too much energy." Mikey was three, and Emily had warned Aaron about how hyper he was. Aaron found out that Kevin also hailed from the east side of the state.

"So, you're from Bend. Do you know any of the Coopers?"

"All 30 of them," Kevin laughed. "You must have played ball against some of them."

"That Mitch Cooper almost single-handedly knocked us out of the playoffs when I was a junior."

Kevin thought for a moment. "That kid was good, but his older brother, Cody, was the best Cooper athlete ever. He was two years behind me and was the best varsity player we had."

"Emily tells me you played a year of football at U of O."

Kevin gave Sara a funny look.

"Yeah, I played there a year before I broke my leg in three places. It was a pretty grim experience."

Aaron asked Sara if she'd ever heard of a professor by the name of Jonathon Finley.

"Why do you want to know? Do you know him?"

"No," Aaron answered. "I've only heard of him." Sara grabbed Aaron by the arm.

"Did Emily show you the goatie?"

"Please, Sara, he doesn't need to see it."

"Emily, everybody needs to see the goatie." The goatie was a small wooden shed with a tin roof. It looked like it had been built in the 1930's by kids who weren't very good carpenters. It came with the property, and it was rumored to have at one time housed goats, though that was never confirmed. When Sara and Emily were growing up, it was their special place.

"We don't need to go inside," Sara said. "Now tell me, why did you ask about the good professor?"

"Why is it a big deal? And why are we talking in private?"

"Aaron, I think that guy is one evil son of a bitch. I had him for a class my freshman year, and he all but transformed freshman into America-haters."

"Hey, Emily", Aaron yelled. "Come check this out." Emily walked the 40 yards to the small wooden shed.

"What's going on?"

"Tell her, Sara."

"Aaron wanted to know about Jonathon Finley. I just told him that Professor Finley was an evil son of a bitch."

"Tell us why, Sara."

"What in the hell do you guys know?" Sara asked. Aaron told Sara about his mother's journal and the meeting of the two professors. Sara listened in disbelief. "That's the strangest thing I've ever heard. What do you think he is?"

"We don't know," Emily continued. "What do you think he is?"

"Someone who doesn't like where he lives, that much I know," Sara explained. "When he talked about our government, there was only one party. The GOP didn't exist, according to him. He said that any true American would do anything in his or her power to destroy conservative values. He preached freedom of speech, and said there were no rules. He said that every corporation in the U.S. was corrupt, and should be put out of business. The strange thing is that most of the students were convinced he was right. He said the Second Amendment was meant for a different time, and that no citizen should have a firearm. He said that it was a woman's right to choose, which is the only thing he preached that I ever agreed with. I could go on and on."

"Don't stop," Aaron pleaded. "This is amazing."

Emily pointed toward the house. "Dad's waving. Let's talk after dinner."

" Okay," Sara said. "There's a lot more you should know."

The dinner lasted an hour, and afterward everyone was moving a little more slowly, except Mikey. Mikey took a liking to Aaron and wouldn't leave him alone. Aaron didn't mind.

Emily grabbed Aaron and took him around to the front of the house. "Sara is going to see if Mom and Dad will watch Mikey for a while. Do you want to go downtown with Sara and Kevin?"

"I don't know, Emily; it's the big city and all."

"Very funny. I'm sure you'll survive." Aaron thanked Don and Trish for dinner, prompting the others to do the same. They piled in the Suburban and just drove around for a while, showing Aaron the view from Council Crest. Sara was the first to mention their previous conversation.

"Where did I leave off?"

"Off what?" asked Kevin.

"I was telling Emily and Aaron some little ditties about Professor Jonathon Finley. Remember him?"

"Only what you've told me," Kevin replied. Sara thought for a moment.

"I remember what he said about being a lawyer. He said that any lawyer worth anything could make a living by inventing a lawsuit. He cited the cases against big tobacco, and told the pre-law kids that alcohol was just ripe for the picking. He mentioned things like mold on lumber that could cause disease. Said it didn't matter if it really caused a problem, as long as you could find enough expert witnesses. He even went so far as to make a case against fast food chains for serving fatty foods. It was really funny because all these pre-law kids were asking questions and taking notes. They were already spending the money from all the frivolous lawsuits."

"What other professions did he promote?" Emily asked.

"Yours was the other biggie. He had some great advice for future teachers. He said that as important as education was, getting compensated for teaching was even more important. He helped form the teachers' union and structure the benefit packages. He said that teachers deserve to retire with more than any other profession because they put up with so much shit."

Emily chimed in, "You know, he may have had some valid points. Why don't we go to Mini Brews and have a beer?"

The music was loud, but Sara spoke over the other sounds and went on to tell the others more about Finley. She could see it bothered Aaron when she told them that the professor was a staunch environmentalist. She told them that his motto was "Do what's right at any cost."

Aaron couldn't help himself, "Like burn down a ranger station or two?" Emily gave Aaron a look, but Sara and Kevin hadn't really caught what he had said. They drank a couple of beers and left to go back to Emily's parents' place. They sat in the driveway for a moment, and Sara offered one last bit of information.

"You probably can't tell, Aaron, but I'm six years older than Emily. I remember the Gulf War like it was yesterday. It was '91, and that was the same year I took the political science class from Finley. He said that the president should be charged with treason for protecting Kuwait. I know it was only because Bush was a Republican. He even helped organize a war protest on campus. It was really funny, because even most of the little liberal shits liked the war." Sara went into the house and got Mikey.

Aaron and Emily said goodbye to Kevin and went inside. On the way out the door, Mikey gave Aaron a high five and gave Emily a hug. Emily and Aaron said hello to Don and Trish, and went out into the backyard. There was a warm breeze, and the sky was unusually clear for the valley.

They sat on the lawn swing and talked. "Why are you and Sara so different?"

"You really think we're different?"

"Well, I would say that she's a conservative, and you're not."

"You sound like my dad. She is a hospital administrator, and I'm going to be a teacher. That's the big difference."

"Do you think she's right about teachers? I mean, do you think they should put their personal interest in front of educating their students?"

"She didn't say that that's what they did, Aaron. She said that is what the professor suggested they do. I'm sure there are teachers who care a lot more about their own welfare than they do about educating kids, but I would think most of them are in it for the kids."

"Nancy was a teacher for quite a few years, and she thinks the system is pretty bad. She thinks that they should make teachers responsible for what the kids learn. But in the same breath, she says that the teachers should have control of how kids act. What grade are you interested in teaching?" Aaron asked.

"I want to help mold the young minds of first and second graders." Emily said.

They sat and talked for over an hour. There was still a warm breeze, even though it was after ten. Aaron finally put his arm around her and kissed her. Emily's hand went right to the inside of Aaron's thigh and moved slowly up and down until she found his buttons. This time, she undid all of the buttons and put her hand on him and began gently massaging him. He reached around her back, and with one quick movement, undid her bra. She whispered into his ear, "You're good at that. I think Mary was wrong."

"Please don't mention Mary at a time like this, and what was she wrong about?" Aaron asked.

"Nothing. Sorry I said anything."

"So is there someplace we can go?"

"Wait here a minute." Emily went into the garage and came back with a sleeping bag tucked under her arm. She took Aaron's hand, and they walked to the little wooden shed. She opened the door very slowly, so it wouldn't creak. Emily spread the double bag out on the ground and laid down.

Aaron fumbled for his wallet and was in the process of retrieving the condom he'd been carrying forever when she stopped him. "You don't need that, unless you have some horrible disease or something. I started taking birth control pills as soon as I got home from the timber tour. I knew this would happen." Aaron sat on the edge of the sleeping bag and took his shoes off. There were spaces in the walls, and the moonlight filtered in and cast a dim light into the building.

Emily took her clothes off first and lay on her stomach. Aaron took his off and started rubbing her back, moving his hands slowly over her entire body. She turned onto her back and whispered into his ear. "I'm ready. I've never been this turned on in my whole life." They made love for the first time there in the little wooden shed. It was getting chilly when they went inside; they were surprised that it was after one in the morning.

Emily whispered into his ear. "Mary had told me you were a virgin. Was she right?" In the dim light of the back porch, she could see his face turning red and was sorry she had said anything.

"Mary doesn't know everything about me." That was the last time the subject ever came up.

Emily was up before Aaron. She went to the couch and kissed him on the forehead. "I had an amazing dream last night, and you were in it."

"It wasn't a dream, Emily."

"Are you sure?"

"Trust me. There are just some things that you know are real, and last night was one of them." Emily put some toast in the toaster and started a pot of coffee. "That's kinda funny, you getting up before me," he said.

"I couldn't sleep. I kept thinking about you and me and last night. It was wonderful."

"Please talk quietly. I'd die if your parents heard us."

"Don't worry. They're on their Sunday morning ritual."

"Church?"

"No, they go for a drive every Sunday morning. They have been doing it for years."

"So how long will they be gone?"

"That's the problem. You never know how long they will be." Aaron grabbed her and pulled her on top of him. He was busy with her bra strap when they heard the car door slam. Emily jumped up, and Aaron quickly pulled his pants on. Don and Trish walked into the front room with a bag of fresh doughnuts. Trish set the bag on the kitchen counter, and gave Emily a funny look as she passed.

Aaron took a shower after breakfast and helped Emily clean up the kitchen. "So what's on tap for today?" Aaron asked. "It'll be hard to top yesterday. That has to rank right up there with first deer and first touchdown."

"Gee, Aaron. It's great being ranked right up there with first deer."

"I was talking about the fishing trip. How about a little trip to Eugene? I think it might be a good idea to go see your old friend Herb." Emily agreed, and they were on the road by nine.

They pulled up in front of Herb's at 10:30, and Emily started laughing.

"What's so funny?" Aaron asked.

"That." She pointed to Rachael's VW parked in the driveway.

"Is that Rachael's?"

"Yes, it is; she is such a liar." Aaron knocked on the door, and they waited for a couple of minutes before knocking again. This time, he made sure someone heard. Finally, Herb stumbled to the door and peeked out. He opened the door and let them in. Herb had on jeans and a tee shirt, and in his right hand he held a 357 Magnum.

"Christ's sake, Herb, expecting company?"

"Sorry, Aaron." Herb put the pistol on the coffee table and sat on the couch. Rachael came out of the bedroom wearing an oversized tee shirt and a big grin.

"Hi, guys. Guess this doesn't look good, but we're just friends."

"That's pretty funny, Rachael. What kind of friends?" asked Emily.

"Really good friends."

They talked about their problems and about what had happened to Ernie Roper. Herb's theory was that Ernie just knew too much. He knew who the head FF members were, and that was information that was just too important to have leaked. Aaron asked Herb about Professor Finley, and Herb told them that the professor had been a major influence on his life. The more Aaron and Emily quizzed him, the more he revealed about the professor. Herb told them that the professor had helped educate such distinguished officials as U.S. senators, representatives, and two governors. Aaron and Emily could hear Herb's love of the professor in his voice.

After a few minutes of exuberantly profiling Professor Finley, Herb abruptly stopped his conversation. "Why do you want to know about Jonathon?"

"Herb," Aaron continued. "I'm going to tell you something, and you're not going to believe me. Please remember that we let you finish your horrific story about the gas cans, so let me finish." Herb sat silently as Aaron began the story of the two professors.

When he got to the part about Professor Finley groping Professor Grayson, Herb stood up and, putting his hands over his ears, shouted, "That is pure bullshit, Aaron! Where in the world did you get that information?"

"Are you going to listen to the whole story or not?"

"Not if it's fictitious, Aaron."

Aaron explained, "This came from a source that would not have any reason to make this stuff up. It was a journal that my mom kept during her time in school here. She was some sort of an assistant for Finley. This information came about by accident. Can I finish this, and then let you decide if it's true or not?"

Herb stood silent for a minute or two, and then sat back down next to Rachael. "Go ahead, but I'm sure this information is false." Aaron finished the story and waited for a response from Herb. Herb sat, contemplating the story for a few minutes, and finally spoke. "If that's true, who is he, and what is he?"

Emily said, "That's what we're trying to figure out. We're sure this happened, but have no idea why."

"Well," said Herb. "If this is true, I'm one stupid son of a bitch."

Rachael and Herb got dressed while Aaron and Emily waited in the front room. It was after noon, and Aaron was getting a bit hungry. He yelled toward the bedroom. "You guys up for some lunch?"

Finally, a reply came from the bedroom. "We'll be ready in a minute."

They took Herb's car to some earthy pizza parlor that served only whole-wheat pizza crust. Aaron wasn't wild about the idea, but decided to give it a chance. Herb ordered a large veggie combo, and Aaron couldn't help himself. "We need some protein on that thing."

" Okay, okay," said Herb. "Make it half-veggie and half-meat-lover's. That work all right?"

"That works great," said Aaron. They quietly ate their pizza and discussed the Finley story. They agreed to meet in Bend for the first group meeting on the first Saturday in July. Herb promised to bring all of the information he had on Finley, and Aaron promised to bring the journal.

Herb and Aaron were striking up a friendship, and Aaron felt weird about it. He knew he detested the things that Herb had done, but somehow, he felt sorry for him. After they were finished eating, they went back to Herb's.

They said their goodbyes, and Aaron and Emily headed back to Tigard. It was after five when they got to Emily's parents' place. Aaron was tired, and Emily could tell. "You better get started. You look beat."

"I am. That fishing really took it out of me."

"Are you sure it was the fishing? You know, I thought good conservative kids like yourself were into abstinence. I'm glad you're not, but aren't you supposed to be?"

"Well, I have a theory about that."

"I've got to hear this. A theory, huh?"

Aaron explained his theory. "I think all the good old boys, be it a minister or a politician, either screwed around, or really wanted to. The ones that really wanted to, but didn't get to, are the ones that make up the rules."

Emily was laughing so hard she had trouble catching her breath. "Did you come up with that on your own?"

"Pretty good, isn't it?"

"Pretty good? I think you hit the nail right on the head." Don and Trish weren't at home.

"Tell your parents thanks for everything, especially for having such a great daughter." They kissed for several minutes, and Aaron finally got into his truck and backed slowly down the driveway. "I love you, Emily."

"I love you, too."

Chapter VIII

Tall cottonwoods and birch shaded the road into the small cabin. Professor Jonathon Finley had owned the cabin since 1965. In the 35-plus years he had owned the cabin, he had visited often. He had spent entire summers there in the forest, just west of Coos Bay. The trip from Eugene was particularly tiring for the 74-year-old professor, especially with so much on his mind. The meeting in Eugene had taken a strange twist, and members of the different pro-action groups were at odds. Friends of the Forest leader Tom Erickson challenged the professor's motives, saying that what was once a noble and just cause had turned into pure political posturing. The one time state senator was one of the professor's prize students, and this was the first time he, or any of the other Leaders of Freedom, had challenged him. The group of five former students, some now in their fifties, had laid a carefully planned ambush for the old professor, and he had fallen victim to it.

Other than Friends of the Forest, there were The Keepers of the Beast, Liberty for America, Corporate Crushers, and the professor's favorite, CSR, which stood for Complete Social Reform. The only underground entity was FF, but all of the leaders were underground. The only people who knew the leaders' identities were the second-in-command of each group, the other group leaders, and the professor. Animal rights activist Patty Marks was the only female leader among the group. Their mission was the ethical treatment of animals, concentrating mainly on research animals. Their group took no responsibility for the fires and vandalism at the different research centers around the state, but they were solely responsible. They also led the campaign to stop predator hunting with hounds, not to save the predators, but to deplete the game herds for hunters. No game, no need for guns. Liberty for America leader Loren James was a brilliant attorney, and a senior partner in the largest firm in the state. His mission was to interpret the Constitution and insure civil liberties at any cost. He was intent on destroying the morals of every student in every school in the state. Strict adherence to the separation of church and state was usually his main focus, but the right to complete privacy also occupied much of his time. Drug testing of student athletes was his current main focus. CC leader Mark Driggers was really nothing more than the leader of every union in the state. His main focus was education. By compensating teachers with low wages and high benefits, he had helped cripple the educational system in the state. His main focuses were extra special education teachers and more administrators than were needed. He also worked with the trade unions, and did everything possible to run every business out of the state. CSR leader Marty Smith was another of the professor's prize students. He was the most successful of all of the leaders, now in his second term as governor. His motive for social change was more of a power trip than true compassion, but those in the metropolitan areas were none the wiser. His claim to fame had been socialized medicine, nearly bankrupting the state. Land use rights were another of his prizes. Making the state the guardian of all lands, public and private, was his main goal.

    The professor had organized the different groups in the early 1970's, and the leaders had been the same since their inception. The leaders rarely communicated with each other unless it was their annual meeting, but prior to this meeting, all five had gotten together without the professor present. This special session, called by the governor, was to discuss none other than Jonathon Finley. In all the years that the groups had been active, there was never any personal injury to a human being, and now there was a homicide to deal with. Ernie Roper had fallen on hard times financially, and sought to extort money from FF. The big problem with Ernie was that he knew who all the leaders were. There had been no other means to deal with Ernie. Professor Finley had suggested that Ernie needed to be silenced, and his plan was foolproof. Get Ernie drunk, and throw him in the river. The plan had worked, but now the game they played had higher stakes. The secret meeting was called to question the goals and methods of the various groups. The big question was why the professor was so intent on destroying what made the state what it was. They started by trying to rationalize his motive for eliminating all of the timber jobs in the state and country. What had once seemed like a good political move had turned ugly. Timber, at one time, helped finance education, and now that money was gone. Ideas like those of the professor had eaten into the state resources, and now, even the most radical environmentalist had second thoughts. Animal rights, while always a good political move, had started to turn ugly, with the numerous bear and cougar encounters being recorded. Patty Marks joked about loving animals, but being afraid to go into the woods. Loren James had forgotten why there could be no Christmas in school. He felt guilty about most of what he had done. Mark Driggers had spent a good portion of his adult life pitting worker against management, teacher against school board, and yet he didn't really know why. The conflict would always exist, but why was he so intent on fanning the flames of unrest? Marty Smith was still proud of his social reform and land use restrictions, but even he could see the heavy financial burden it had placed on the state.

The research had taken several days, but, aided by special government perks, it was accurate and complete. The immigration records had been forwarded only two days before the scheduled leaders' meeting, and the governor had called a second secret meeting to discuss their plan. It had been found that the Finleys, while immigrating to the U.S. legally, had no record of life in Ireland. The couple had entered the States through Ireland, but were not of Irish origin. Information about the professor's brother had also turned up disturbing information. It seemed that the professor's brother, Ned Finley, had taken the same path as the professor. The only difference was that Ned Finley had committed suicide in 1993 after being questioned by the FBI. It seemed that some of his law students at Harvard had gone to the FBI because they viewed his ideas as damaging to the security of the country. The FBI was unable to resolve where the Finleys had come from, but was able to find that, along with the Finleys, there were nine other couples who had entered through Ireland, but had never lived there. The scheduled meeting had taken place in the same manner as it had for the past 27 years, but this time the professor was questioned about his motives. He tried to convince the five leaders that everything he had done was to benefit America, but he knew he hadn't convinced them. Governor Smith was the only one to come to the professor's aid. Arguments took place among the different leaders. The professor excused himself and left the meeting early.

The light in the cabin was on, and Professor Finley knew that Amy was already there. The only person with whom he had ever been intimate was waiting for him. "It's good to see you, Amy. How was your trip?"

"What's wrong, Jonathon? You don't look well."

"They know, Amy. They know I'm not who they thought I was. They are suspicious of why I have pushed them all these years to see things my way. You should leave. I don't want you implicated, in case they go to the authorities."

"Think about that statement! For Christ's sake, Jonathon, they are the authorities."

"That they are, and after 30 short years, they caught on. Smith is the only one who still thinks he's a saint. The others know they were grossly misled. All that time spent pushing and pulling them down the path of deception, and one stupid mistake, the murder of Ernie Roper, changes the course of history."

"Do you really care? No one can hurt what family we had, and no one can hurt us. We know too much."

"Amy, you are the only bright spot in my dismal life. I'm glad we were able to have had the time we did. It was fate that sat us next to each other so many years ago. If there is one thing in life that I am grateful for, it is you."

"Jonathon, do you ever feel guilty for doing so much damage? You've done such a good job of turning this state upside down."

"It was what we were supposed to do. Do you ever feel guilty for altering the entertainment industry? They lean so far left that they keep falling over, and don't even know it. And what about the media? It's almost funny how stupid they've become. Only trouble is, people are catching on."

"You're right, Jonathon, but who could have ever envisioned talk radio becoming what it is today? It was one factor we hadn't plugged into the equation."

"Amy, do you think they did it again?"

"Did what again?"

"Do you think there was another batch of us, or maybe batches, or do you think we were it?"

"I don't think Stalin would have wasted time on another batch, but if he had had any idea how effective we were, there would have been thousands of us."

"You're right. There would have been thousands."

Roderick Larios

Chapter IX

Jed and Aaron walked a good 30 miles the last week of June. They surveyed, cruised, and marked over 150 acres of federal timber. The Trinity Butte sale was one of five awarded to mills in Eastern Oregon. There had been few sales over the past 10 years, and most of the ones that were awarded were tied up in court battles with environmental groups. The Forest Service had assured A.J. that this sale would be immune to litigation under the "Federal Forest Fire Prevention Act." Douglas Land and Timber had been the only mill to bid on the sale, and it was the only sale that permitted clear cutting. Several environmental groups were outspoken opponents of the sale. The only group that wasn't involved in this particular protest was none other than Friends of the Forest.

Most critical of the sale was The Cascade Club, with all of its high-profile members and left-wing politicians. There were several motions to stop the sale, but not even the most liberal judges would issue a court order to stop it. The sale was a sealed bid, and Douglas Timber got it for a song. The bid price was leaked not three days after the sale had closed, making the radical environmental groups even madder. A.J. was confident that the logging would go smoothly, but took some precautions just the same. The night watch was beefed up at the mill, and each logging site had someone on hand 24 hours a day.

The sale lay between a wilderness area and a federal track of land that had not been harvested for over 80 years. The federal track was heavily laden with fuel that was tinder dry. The track to be logged was a long, narrow strip of land, approximately 300 feet wide and a little over four miles long. The sale followed the terrain and, at some points, crossed small portions of the wilderness area. At noon on Friday, Aaron and Jed had completed the final prep work for the project. Early Monday, the logging company would be on the job.

On their way through town, Aaron spotted something and asked Jed to turn the rig around. "What is it, boy? A good-looking girl?"

"Not exactly. You're gonna shit when you see this." As Jed rounded the corner by the Motel Six, he saw a sight that made him turn three shades of red.

"What do you think those crazy bastards are here for?"

"Oh gee, Jed, let me see. They're probably here on some sort of a fact-finding mission. Maybe they're doing a study on the area to see if it's suitable for a commune."

"Listen, you little shit, don't be a smart ass. I know what they're here for."

"Sorry, Jed."

Three longhaired, hippie-looking guys and four girls were getting out of a VW bus. They wore dirty clothes and carried ragged-looking backpacks. A "Save the Earth" sticker adorned the back window of the bus.

"How many of those idiots do you think will show up?" Jed mused.

Aaron considered the question. "No idea, but any are too many."

"You're right about that. Let's go see what your old man has to say about this little situation." They drove through town and toward the mill, passing another VW bus full of the same type of people. "I hope the Forest Service keeps their promise about prosecuting anyone trying to obstruct this sale. If they let these dipshits stand in our way, it'll sure screw up our log supply this summer."

"I wouldn't worry too much, Jed. Dad says they're adamant about getting these fire buffer zones before the fire season hits."

A.J. was in his office and wasn't pleased with the news about the tree-huggers. "Bill Adams assured me that anyone trying to stand in the way of this cut would be shackled and carted off. Bill might work for the government, but he's still an honest guy. I'm not too worried about this. I have a word of caution, and spread the word. If there is someone sitting up in a tree, do not cut the tree down." Aaron had to laugh at that.

"You're no fun at all, Dad."

"You never know, Aaron. Some of those radical little shits might be pretty nice."

Aaron thought of his mom and then of Emily. "You're probably right, Dad."

Aaron called Emily as he drove toward the ranch.

"Hi, Aaron! How's my favorite tree killer?"

"Fine, and how's my favorite swimming instructor?"

"Haven't drowned anybody yet." Aaron told Emily about the new group in town and about the sale. She begged him to be careful, and then told him to stay away from all of the tree-hugger girls, because of his uncontrollable attraction to them. Aaron assured her that he'd keep his distance.

They talked until he pulled into the driveway at the ranch. "I'm home. I can't wait until next week."

"Neither can I. I love you."

"I love you, too."

Mary ran out of the house with the look of someone who had just won the lottery. "Aaron, you're not going to believe what I found."

"Try me. I believe most anything."

"Have you ever heard of a tabloid called *Believer*?"

"Are you talking about one of those gossip newspapers, with all the weird stuff in it? Like aliens and all that crap?"

"Apparently, these things have been around for a long time, this one since the '40's." Mary showed Aaron a copy of a story that she had pulled up on the Internet. The headline read, "Russian Couple Slain after Revealing Communist Plot." The story told of a young Russian couple that had been forced to marry, and were relocated in the United States. It told of their families being tortured and killed in Russia because of their unwillingness to carry out the mission. The story told of some 200 couples that had been sent to various countries to have children, and to educate their offspring in order to disrupt the culture of the various countries. Before the authorities could investigate, all of the main players were dead, including the reporter that had garnered the story. The part that Mary was the most excited about was the names of four couples that had supposedly been included in the group sent to the United States. Aaron's mouthed dropped as he read the names. The second couple mentioned was none other than a Jake and Rebecca Finley.

"Holy shit, Mary! This story is true."

"Not according to the *Washington Post*. There were several articles written about the story, and all of them came up with the same conclusion. They said the story was a big hoax, and that the couple was from Iowa, and made the story up so they could sell it to the *Believer*."

"What do you think?"

Mary didn't hesitate. "I think we found out why Professor Finley is such an asshole."

Social Manipulation

They agreed to keep the information to themselves, at least until after Aaron had a chance to talk to Herb about the situation. Aaron told Mary about the new people in town, but she already knew.

The word was out that the timber sale on Trinity Butte would, under no circumstances, take place. Mary explained to Aaron, "It was on the Portland News at noon, and showed about 30 long-haired guys and ugly girls getting ready to wage war with the Forest Service. What does Dad say about it?"

"Not much. He just told us not to cut any trees down if there were people in them. He also said something that I thought was a little strange. He said that some of the radical little shits were probably pretty nice people."

"Why do you find that strange? The girl he married was one of those people." Aaron took the opportunity to tell Mary the whole truth about Emily and Rachael, and about FF and the fire, and pretty much everything else he could think of. Mary sat there, stone-faced, as if she were in shock.

" Are you all right, Mary? I didn't mean to scare you. I just thought you should know the whole truth."

"Do me a favor, Aaron. Never tell Dad that."

Aaron's job over the weekend was to help guard the equipment that was in place up on the Trinity Butte. They were to work in teams and under no circumstances come into contact with any of the protesters. There were to be no weapons. Friday afternoon, Aaron stuffed his Hi-Standard model-B pistol into his backpack and threw it behind the seat of his truck. He stopped by and picked up Hank Barnett. It was just before six, and the drive to the butte would take almost an hour.

"Hi, Hank. Have you eaten yet?"

"Yeah, but I'd like to stop by the Chevron and get something to munch on."

"Me, too." They picked up a few rations and proceeded to the equipment up on the butte. It was decided that they would take shifts, two hours on and two sleeping, until their relief came at 8:00 Saturday morning. As they pulled into the landing, they saw Eddie Bridges and Art Miller standing in the back of Eddie's pickup, looking through a pair of field glasses.

"Tell me you're looking at the biggest bull you've ever seen." Aaron waited for a reply. "What is it, Eddie?"

"Ain't no fucking elk, that's for sure. Jump up here, and take a look at this shit." Eddie handed Aaron the glasses and pointed in the general direction of the subject. About 500 yards down a draw, there was a small clearing. At the base of a huge Ponderosa Pine stood a blonde girl wearing coveralls and carrying what appeared to be a sleeping bag. About 50 feet up the tree was a small platform that extended about four feet out from the tree. Sitting on the platform was another person.

Aaron couldn't tell if it was a girl or guy, but whatever it was, it was the wrong tree. "That's pretty funny. They're 100 yards from the sale. Look at the markers along the west edge."

Eddie was quick to reply. "Why don't we go down there and help them find their way to the right bunch of trees?"

"Stay away from them, and any other people you see hanging around in the trees," Aaron went on. "The worst thing we can do is get in a pissing match with these people."

Eddie and Art took off, promising to be back at 8 A.M. sharp. Aaron and Hank sat and talked until the sun went down. Just before the last bit of light left the clearing, Aaron took another look at the big Ponderosa. The people were gone, but the platform was still there, and still in the wrong tree.

Eddie and Art pulled in just before nine. Eddie came up with the flat tire story, but Aaron and Hank knew Eddie too well. Eddie had his truck positioned so that he could get a good look at the clearing. "Anything exciting happen last night?"

No one answered for a moment. Finally, Hank spoke. "It was great, Eddie. They were down there screwing their brains out, and didn't have a clue we were here." Eddie grabbed the field glasses out of the cab and quickly looked toward the clearing.

"Where did they go?" asked Eddie.

Aaron leveled with Eddie. "They left the same time as you guys did last night. If they come back, don't go down there."

"Don't worry," Art said. "We wouldn't waste our time."

The first thing Aaron did when he got home was to take a nap. After a couple of hours of real sleep, he felt like a new man. The good thing about guarding the equipment was not having to work in the mill, something Aaron often did on weekends. After lunch, he called Emily, and they talked for 15 minutes. Emily had arranged the meeting for the following week and filled Aaron in on all the details. Herb, James, and Rachael would all be there, with all the information they had about the professor. Aaron told Emily about the article that Mary had found, and asked her not to tell anyone about it until the meeting. After they were done talking, Aaron drove in to the mill to fill A.J. in on the tree people.

"In the wrong tree, huh? That might be the funniest thing I've ever heard." Aaron was glad A.J. liked the story, but knew that his dad was worried. The log supply was something that they counted on. If they had to scrap the Trinity Butte sale, it would mess the whole mill up for at least a couple of weeks.

"I wonder why people do stuff like this? You'd think they had better things to do," A.J. said.

"I don't know, Dad. Did you ever ask Mom why she did this stuff?"

"I asked her, but never really got an answer." A.J. looked at him. "Did you ever ask Emily why she did this stuff?"

Aaron started to shake. All he could think about was kicking the shit out of his little sister.

"How do you know about Emily? What did Mary tell you?"

"Mary didn't say a word to me. Mark Herman saw you and those two girls at the mill picking up the spike pullers. Didn't take a rocket scientist to figure out what you were doing."

"Why didn't you say something before this?"

"I wanted to see if you'd tell me the truth."

"Well I wouldn't have, not in a hundred years."

"Tell me, Aaron. Would I be correct to assume that your friend Emily has given up the practice of tree spiking?"

"That would be very correct."

"Good. Let's make this the last conversation about this. I don't like you to lie to me, or to anyone else, but I understand why you did it." Aaron left his dad's office feeling about as low as he had ever felt. The thing he hated the most was disappointing A.J.

Aaron picked Hank up at the same time as the day before and headed for the Chevron station. They got to the equipment site about seven, and there was Eddie, standing in the back of his truck, staring through the field glasses.

"I take it they're back."

"Oh yeah, and they're both girls, and they're both kinda pretty."

"They're both kinda stupid," Aaron added. "Sitting in the wrong tree and all." Aaron asked Eddie if he'd mind leaving the field glasses for the night. As Eddie and Art pulled away, a strange feeling came over Aaron, like something bad was going to happen. He thought for a moment and realized he was just upset about the conversation he'd had with his dad. Surely that was it.

Social Manipulation

Aaron and Hank talked until dark, sitting in lawn chairs in the bed of Aaron's truck. Just before dark, Aaron took a last look at the two girls. One was on the tree platform, and the other was curled up in a sleeping bag at the base of the tree. It was about 10 P.M. when Hank went to sleep. Aaron could see by the light of a half moon, and soon his eyes became adjusted to the dark. Moonbeams cast an eerie light through the trees, and Aaron kept thinking that he saw movement toward the clearing.

Just before midnight, Aaron gently shook Hank. "Your turn, partner."

"Go away, man. I'm really comfortable."

Aaron was just about to make a smart-ass remark to Hank when the screaming started.

"What the hell's going on?" Hank yelled.

"Let's go." Aaron grabbed his pistol out of his backpack and the big flashlight from under his seat. Hank had his shoes tied just as Aaron slammed the pickup door. They ran through the trees and across a small streambed. Hank had trouble seeing and tripped over a log.

"You all right Hank?"

"Yeah, it's just a flesh wound." They were about 40 yards from the tree, and the screaming was getting louder. As they approached the tree, they could see the girl in the sleeping bag. Her neck was bleeding, and her face was bloody and very dirty, like someone had rubbed her face in the dirt.

"What happened?" Hank yelled at the girl. She couldn't talk. She just pointed upward, toward the tree stand. As Aaron turned his flashlight upward, he could see the green eyes reflecting in the light.

"Holy shit, Hank, it's a cat! Quick, hold this light for me." Hank shined the light on the cougar. It was about five feet below the girl in the tree, and she was screaming at the top of her lungs. Aaron had to yell as loud as he could to be heard over all the commotion. "Move around the tree, away from the light." The cougar was about to jump toward the girl as Aaron started firing. He saw a branch shatter with his first shot. The second, third, and fourth shots found their way through the cougar's rib cage and into its vital organs. Aaron emptied the entire clip as the cougar tried to make one last desperate lunge toward the girl. It fell like a pinball between the branches and landed about a foot from the girl in the sleeping bag. She had stopped screaming and was sitting there in shock.

The girl in the tree hadn't stopped screaming, and Aaron yelled at her to please shut up. Hank held the light as Aaron climbed the tree to where the girl was. He reached his hand out, and she grabbed it and climbed back on to the platform. She was shaking so hard that she could barely talk. "It tried to kill us, didn't it?"

"It tried to have you for dinner," Aaron explained. "But it didn't. You're okay." Aaron helped her out of the tree and sat her by her friend. He examined the bleeding girl as Hank held the flashlight.

"We had better get you to town. Where's your rig?"

"We don't have a rig," the girl from the tree explained. "We were dropped off here."

"Where's the sorry son of a bitch that dropped you off?"

"Don't know. He said he'd be back in the morning to check on us."

"Let's go." Aaron helped the bleeding girl to his truck and grabbed his first aid kit from under the seat. He applied a bandage to her neck, trying to stop the bleeding.

They all got into the front of the truck, and Aaron headed toward town. "How did you get the cougar off you?" Hank asked the bleeding girl.

"I climbed back into my bag when it let go of me for a second."

"Smart girl," Hank replied.

The ER doctor put 32 stitches in the girl's neck and three over her eye. The wound over the eye was neither a fang nor a claw wound, but apparently had been caused by a rock or other sharp object. The cougar had pushed her head into the ground with enough force to cause the wound. She hadn't lost enough blood to warrant a transfusion, but the ER doctor wanted to keep her in the hospital for observation. The girls explained to Aaron about their mission. They were to stay in the tree, no matter what. They would prevent the cutting, and therefore save the forest. Hank was so mad that he left the hospital and walked home.

As he was leaving, Aaron asked him if he'd drive back up to the logging site. "Might as well," Hank replied. "I'd never get to sleep after all this."

Aaron found out that the girl in the tree's name was Sarah, and the girl who was attacked was named Ann. Aaron let Ann use his cell phone to call her parents. She hung up when her mom answered the phone. "I better wait until morning. My parents are going to freak out. When my dad finds out what I came up here for, he's going to kill me."

Aaron couldn't resist. "Who told you what tree to sit in?"

Sarah answered, "Our team leader, Gary Wright. Why?"

"The tree you were in was over 100 yards away from the sale. You risked your lives saving a tree that wasn't even going to be cut. If we hadn't been up there guarding our equipment, you'd both have been cougar chow."

"I'm sorry." Sarah was crying. "I never even thanked you. What were the chances that there would be someone close enough to save us?"

Ann interrupted. "I don't know what to say. I've never had my life saved before. If there is anything I can do to repay you, just let me know."

"Promise me that you'll give up tree sitting."

"Don't worry. I'm through saving the forest."

When Aaron got to the ranch, he went straight to A.J. and Nancy's bedroom. "Dad, I need to talk to you."

"What's the matter, Aaron?" A.J. got out of bed and went out into the hallway. He motioned Aaron to follow, and they went downstairs. "What happened? Is anybody hurt?"

"As a matter of fact, I just came from the hospital."

"Jesus Christ, Aaron! I told you not to get into a fight with these guys."

"It wasn't a fight. It was a cougar." Aaron told A.J. the story. A.J.'s eyes were almost as big as the cougar's had been. The longer the story went on, the more engrossed A.J. became.

When Aaron finished, A.J. sat there for a minute and didn't say a word. "You're not kidding about this, are you?"

"No, Dad, that's just what happened. If Hank and I wouldn't have been out there guarding the equipment, I'm pretty sure one or both of those girls would be dead."

"Why did you take the pea shooter? I said no guns."

"It was behind the seat in my pack, just for safety."

"Thank God you don't listen to everything I say."

By the time Aaron and A.J. were done talking, the first rays of sunlight peeked through the big picture window. Aaron wasn't thinking about sleep; he was thinking about the guy who had left the two girls in the woods. A.J. went back to bed.

Aaron used his cell phone to call the sheriff as he headed back to town. His first stop was Eddie Bridges' place. They took Eddie's truck and stopped by and picked up Art. Aaron told them the story as they drove to the Motel Six. Both the VW buses were parked in the lot.

"How you gonna find out what room the jerk's in?"

"Watch and learn, Eddie." Aaron went to the front desk and rang the night bell. A lady who looked to be in her thirties, and very tired, came through the door that was behind the counter.

"What can I do for you, gentlemen?"

Aaron answered, "We need to speak to Gary Wright. He left two young girls out in the woods last night, and a cougar attacked them. We're here to kick his ass." The lady stood there for a second, grabbed the guest book, and scrolled down the page with her finger. "You better not be lying. He's in Room 112, just around the corner."

A skinny girl with long black hair answered the door. Aaron pushed his way through the door as the guy in the bed scrambled for the bathroom. Eddie was quick enough to get there before he did.

"Not so fast, asshole. We have some business to discuss." Aaron was standing in the doorway, and Art was right behind him. "Why did you leave those two little girls out there by themselves? They were attacked by a cougar last night, while you were in here just having a ball." The guy was shaking and almost crying when he spoke.

"I think you're full of shit. You better get the hell out of here before I call the cops."

"I already called the cops. They're gonna want some answers." Aaron continued. "How many kids did you deposit out there last night?"

"None of your fucking business." Aaron started for the guy. "Okay, okay. There were two teams, two girls at one point, and a girl and a guy at the other location. You're lying about the cougar. Cougars don't attack people."

Aaron motioned for Eddie to leave as he spoke to Gary Wright for the last time. "Why don't you go over to the hospital on Grant Street and tell Ann they don't attack people? That might make her feel better."

Eddie and Art didn't say anything; they had never seen Aaron so pissed. They drove to the equipment site and found Hank asleep in the back of his truck. As they got out of their truck, a state police rig pulled up behind Eddie's truck.

There was a fish and game officer and a patrolman in the white Chevy pickup. Aaron didn't know either of them. The fish and game guy spoke first.

"So who's been hunting mountain lions without a tag?" Aaron stepped forward and told them the story as they walked to the scene of the attack. The nametag on the fish and game guy's uniform read J.D. Hunt. Aaron thought to himself that the guy's name really fit him. As they got to the base of the tree where the cougar was lying, J.D. Hunt spoke. "It's just a matter of time until someone gets killed by a cougar or a bear. This isn't one of your starving old cats that couldn't catch a deer. This guy was in good condition." He showed the others the wear on the fangs and guessed the cat's age to be four or five years. J.D. Hunt and the patrolman found a pole and tied the cougar's legs around it. They carried it to the landing and put it in the back of the pickup.

"Well, thanks for calling us. We'll split him open and see what was making him tick. I doubt the rabies test will be positive, but we have to check."

Aaron asked a last question. "How many near-misses have there been?"

Officer Hunt answered as he slid into the passenger side of the pickup. "Too many."

Hank dropped Aaron off in front of Eddie's house. "We're off duty for the rest of the weekend, aren't we?"

"Sure are, Hank. Hey, thanks a lot for going back out there last night."

"I had to get out of that hospital. I was about ready to beat the shit out of those stupid girls."

"Wasn't their fault, Hank."

"They were the ones trying to fuck up the sale, Aaron. If it wasn't their fault, whose was it?"

"I'm not really sure yet, but I'll let you know when I find out."

The Forest Service was true to their word. When the logging started on Monday morning, there were 20 uniformed officers waiting to assist the operation. They pulled two people out of trees, a girl and a guy. The row of protestors that tried to block the skid trails was arrested. In a matter of just over an hour, the protest was over. The first logs hit the mill just after noon. Aaron was at the log site all morning, looking for Gary Wright, but he was nowhere to be seen.

In the early afternoon, a fancy-looking luxury car drove to the edge of the landing, and a tall, gray-haired man wearing jeans and an Oakland Raiders tee shirt got out and walked toward the landing. Aaron looked at the car and could see someone through the tinted windows. He thought to himself that this was probably someone from the Cascade Club, stopping by to give them a hard time.

"Where can I find Aaron?" the man asked.

Aaron wasn't really looking forward to a confrontation. The guy was probably in his fifties, but he looked fit, and really big. "I'm Aaron. What can I do for you?"

The guy stepped toward Aaron and stuck out his hand. "You already did it for me, son. You saved my kid's life." Aaron saw Ann step from the car and start toward them. Her neck was bandaged, as was the small wound over her right eye.

She looked like she'd seen better days, but managed a smile when she spoke. "Well, I was wrong. Dad didn't kill me like I thought."

"I can see that," Aaron said. "Considering what you went through, you look pretty good." Ann's dad put his arm around her as he spoke to Aaron.

"It's hard to know what love is until you have a kid of your own. I can't thank you enough for what you did. If there is ever anything I can do for you, please don't hesitate to ask." He handed Aaron a business card, and Aaron put it in his shirt pocket. Ann gave Aaron a big hug and thanked him again. Her dad followed her to the Lexus, and they both waved as they turned around at the edge of the landing and drove off.

Chapter X

The logging sale on Trinity Butte was proceeding without a hitch. By Friday, things were going so well that A.J. let Aaron leave work a couple of hours early.

Aaron had been trying to get a hold of Emily all afternoon, but was never able to make contact. He was on the road by three and tried to call her every few minutes. He pulled into the River Place just before 8 P.M., and the first thing he noticed was Emily's Explorer in the lot. He used his cell to call the motel's front desk, and then asked for Emily's room. He was relieved when he heard her voice.

"What's the matter with your phone?" he asked. "I've been calling for hours."

"I don't know. Maybe I forgot to pay the bill."

"What room are you in?"

" Room 318. It faces the front."

Emily opened the door and pulled Aaron inside. "Hi, handsome. We don't have to meet those guys until ten. How's my real live folk hero? My dad thinks you're Davy Crockett."

"I was amazed about how little press a cougar killing got. It was on TV one time for about two minutes."

"Well, there was a big article in the local liberal press. They even quoted your friend J.D. Hunt on his prophecy. I hope it never happens."

"I hope it doesn't either. It would be a terrible death."

Aaron slipped out of his shoes, and before Emily had a chance to think about it, he had her tee shirt and bra off. They made love until a quarter of ten. Just as Emily was pulling on the last article of clothing, there was a knock on the door. Aaron gathered his clothes and went into the bathroom. Aaron heard Herb's voice say, "Is this a bad time?"

Emily answered, "No, but your timing is a bit off." Herb, Rachael, and James all came through the door and sat at the small table on one side of the room. Aaron came out of the bathroom wearing all of his clothes and a big dead giveaway smile.

Herb stood and reached his hand out to Aaron. "I can't believe you shot a cougar and saved some girls' lives."

"It was nothing. You'd have done the same thing."

"Actually, Aaron, I doubt I'd have the balls to go anywhere near a cougar, but thanks for saying that." Aaron shook James' hand and gave Rachael a big hug. He suggested that they get on with their little meeting, and Herb seconded the idea.

"I've got some info here that is absolutely amazing." Aaron was grinning ear to ear. "But why don't you guys start? What have you found out since we last met?"

Herb spoke first. "Well, first of all, they ruled Ernie's death as an accident. I'd say bullshit to that, but who knows? Next, we think that Finley was intent on disrupting everything that made this state run smoothly. Most of the radical groups in the area followed his ideals. All of the left-wing politicians practiced their governing just as Finley had prescribed. We know that the governor, a U.S. Senator, and at least three U.S. representatives were among his students, not to mention several legislators at the state level. What we don't know is why he did it."

Aaron pondered the question for a minute. "I don't think it's a matter of why. I think it's a matter of who."

James stood and walked toward Aaron. "What is that supposed to mean? We know who we're talking about."

Aaron pulled the copies of the article about the Russian plot out and gave everyone a copy. He sat on the bed next to Emily and waited for them to finish reading. Herb finished first. "Holy shit, this is the strangest thing I've ever seen." Aaron asked Herb to wait until everyone had finished. Herb re-read the article.

The room consisted of four dazed college kids and a cougar-killing-anti-tree-hugger who felt like the smartest guy in the world. "Anybody think there's any merit to this story? My sister Mary found it buried deep in the bowels of our PC. The story has been portrayed as a hoax by the mainstream media, but with the information we have, I think it's more than just a coincidence. Herb, what do you think?"

"I think we're sitting on the story of the century. Do you guys have any idea what this means?"

Emily spoke. "Why don't you tell us? You're the one getting your doctorate."

"Well, Emily, that's true, but a first-grader could figure this shit out. It's all too plain now. The fucking Russians somehow planted their people here, and educated their kids to be professors. It probably wasn't that hard to do."

Herb picked the article up and scanned it again. "Look here. It says they came through several different countries. I always thought Jonathon's little Irish accent was marred. They should have made him a German."

Aaron stood and paced back and forth like a scholarly icon. He finally sat on the edge of the bed and asked the question everyone dreaded. "What do we do now?"

They all went to the Pinewood Brewery and had a few beers. They decided to meet Saturday afternoon at two, back in the motel room. They were to ponder the situation and come up with ideas on how to proceed. They took an oath to tell no one. They all knew that this was a huge story, but were also ever mindful of their involvement in FF.

Emily and Aaron excused themselves at around midnight and went back to the motel. It was the first time that they had spent the night together under normal circumstances, and they made the most of it. They fell asleep in each other's arms, totally naked, totally in love.

Aaron and Emily were ready to roll by seven. They went to a little café on Highway 97, just south of the Burns cutoff. Aaron ate like he was a starved refugee. Emily started laughing.

"What's so funny, Emily? I'm a growing boy."

"I don't think that's it at all. I think you just worked up one hell of an appetite."

"That just might be the truth. I had no idea being in love was so exhausting." They ate their breakfast and drove south on 97.

"Where are we headed?" Emily inquired.

"Seeing as how we have some time to kill, I thought I'd give you some more forestry lessons."

"I believe you. I mean, about the forest. I really don't need another lesson."

"It's not far. I've never actually seen the area we're going to. I just want to see if the information I've heard is true."

"What have you heard?"

"I've heard that the national forests in Central Oregon are a mess, and that private forests are being over-cut. There have been mills that have geared up to take any size logs. It's their only option without getting any federal timber. I just wanted to see what it looked like."

"Promise there won't be any bears or cougars?"

"I promise."

They drove for about an hour, and Aaron found a main logging road that headed east and connected Highway 97 with Highway 31. The first three or four miles were private lands, and the information Aaron had heard was correct. The only trees that were left were scrubby little lodge pole pine trees, and the seedlings that had been replanted. There were no clear-cut areas, but the forest looked barren. The stumps told the story. Trees as small as six inches around were being harvested. About four miles off the main road, they passed the first "Entering Freemont National Forest" sign. Aaron parked his truck, and they got out and walked into the woods. There was patch after patch of large Ponderosa Pines, and the ground was heavily laden with fuel. A mixture of pine needles, branches, and fallen trees covered the forest floor.

They walked a good mile into the forest. Aaron sat on a fallen tree, and Emily joined him. "This is absolutely horrible." Aaron was staring off into the forest. "All this wood, just going to waste. This whole area is a fire waiting to happen; all these trees that could be providing jobs and revenue are just sitting here, waiting to burn. Liberal judges and idiots that think they're protecting the forest are causing more damage than you can imagine."

"Thanks for calling me an idiot."

"I didn't call you an idiot, Emily."

"Yeah, you did, but don't worry about it. You're probably right."

They got back to the truck and kept driving toward Highway 31. The private lands along the way were the same, cut to the bone. And the federal land was overgrown. As they approached Highway 31, they saw a line of firefighting trucks headed south toward Lakeview. "I feel bad about the forest, Aaron. I suppose you're going into the 'See? I told you so!' mode now."

"No, I just can't believe where this whole thing got its roots. Who on earth would have thought that a foreign power would have had such an impact on us?"

"What are we going to do? Maybe I should ask if there is anything we can do."

"First off, I want to clear something up. I think you're the smartest girl I've ever met. You're not an idiot by any means. You were doing what you thought was right, and you can't fault a person for that. I, on the other hand, am an idiot. I think there are a lot of ill-informed people running around, and I'll do what I can to point them in the right direction. I'll never think of you as anything but bright and beautiful."

"Wow, that was a great speech, and if there was an apology thrown in, I accept it."

Aaron pulled his truck to the side of the road and put his arm around Emily. He kissed her and whispered as he pulled his lips away from hers. "I love you so much, and by the way, there is something we can do."

They stopped in La Pine and had a sandwich. In the café, there was talk of the fire 30 miles to the south. Apparently, it had started from a lightning strike late Friday afternoon, and had spread to over 200 acres by early morning. The people in the café seemed indifferent toward the fire, like it was an everyday occurrence. As they got in the truck to leave, they could see smoke to the south, rising high over the mountains. As Emily got into the truck she asked Aaron, "What?"

"What does that mean?"

"What can we do? You said we can do something about Finley and his crew."

"Oh, that. I think we have but one option, and that's to get the information to the federal government. I would be afraid to tell anybody in Oregon what we've found. With all the liberal bias here in this state, they may take his side. That was a dumb thing to say; they already have sided with him. We'll find the right avenue, and we'll spill the beans on old Professor Finley."

They got back to the room just after noon, and Aaron was up for more adult fun, but Emily was still a little mad about Aaron's attitude earlier. Aaron spread all of the material he had onto the bed and tried to organize it the best he could. He put all of the notes he'd taken on one side, and the article from the *Believer* on the other. Everything seemed to fit into place. His mom's journal was the first in a series of events that pointed to the professor being an agent, rather than an immigrant. Then there was the information from Herb and from Emily's sister, Sara. And last but not least, the article from the *Believer*. There was just too much evidence for the whole thing to be a coincidence. Aaron was sure he was right about who and what the professor was.

Emily got up from reading her book to answer the door. Herb and Rachael came in and sat at the table. "Where's James?" asked Aaron. "I hope he's coming to our meeting." Herb looked Aaron in the eye as he spoke.

"I think we have a defector. I think he's going to try and save his ass by going to the police and ratting the rest of us out."

Emily stood next to Aaron with a stunned look on her face. "Oh, this is great", said Aaron. "Do you know this for a fact, or is this just speculation?"

"I'm not a hundred percent, but he just told Rachael and I that we ought to go to the authorities and tell them everything. I'm pretty sure he's gonna spill his guts to whoever will listen."

Aaron gathered his papers from the bed as he spoke. "Is he still here in Bend?"

"I think he is," Rachael said. "He wasn't finished packing when we left."

"Then I suggest we go find him and try to get him to change his mind."

They took Herb's Toyota and got to the motel just as James was pulling out of the lot. Herb signaled for James to pull over, and he did. As Aaron and Herb approached James' car, he rolled the window down just enough to be heard. "Don't try to talk me out of this. I know it's the right thing to do. We should have done it sooner. If we keep playing this stupid little game, we'll wind up just like Ernie." James started to pull ahead, and Emily stepped in front of the car. He put his car in reverse and was unlucky enough to see Rachael standing directly behind him. James shut the motor off and got out of the car. He walked toward Aaron and began to speak. "What do you plan on doing, Aaron? I don't think you have a plan."

"Well, first of all, I'm the only one here whose ass isn't in deep shit. You guys could do time in the slammer. Don't you think we ought to at least talk about it before you spill your guts to the police?"

James walked to a retaining wall on the side of the motel and sat down. His head rested in his hands, and it was hard to tell if he was crying or not. Finally, he lifted his head and stood up. "All right, let's talk. I just need to know where all this is going."

Emily walked over to James and put her arm around him. "Do you want me to ride over to our motel with you? I'll drive."

"Thanks, Emily. I appreciate it."

The meeting started with Herb suggesting they make a pact not to mention anything they knew until it was a group decision. Everyone agreed. Herb had the initial comments. "We know, or at least we think we know, that the whole deal with Finley is an ancient Russian plot from some time in the 1920's. If we believe Aaron's article, we can assume that Finley and a hundred plus like him have been wreaking havoc on the U.S. for nearly 50 years. We don't know if there were other groups, and if there were, we don't know how many. I'm guessing his mission was to mainly educate clones of himself, which he did a really good job of. We need to be very careful of who we talk to, in case we find either sympathizers or other agents. I guess that's the big question. Who do we talk to?"

Aaron stood and took the floor. "You're right, Herb. We do have to be careful. What do you think about trying to get more info on Finley? We could wait until he was gone and search his house; surely there has to be some sort of evidence there."

James spoke next. "I think I know where he lives, and it's out of the way, almost out of town. Two years ago, when they dedicated the building to him, they interviewed him at his house, and I recognized the area. It was pretty funny, come to think of it. He really didn't want to be interviewed, or have a building named after him. It was almost like he was hiding something."

Emily stood and spoke. "It's only a little over two hours away. Let's go." Everyone looked at everyone else, and finally there was a community shrug, like "Why not?"

Herb and Rachael packed up and checked out. Emily left her car and all of her things at the motel. They were on the road by four.

Roderick Larios

Chapter XI

The street looked a little different to James, but he knew they were in the right area. "I think it's the gray house, right there."

"Okay," Emily said. "Wait here." They were parked about 100 yards away from the gray house, on the side of the road. Herb's car was in the front of the procession, with James' car in the middle, and Aaron's truck in the rear. They watched as Emily walked up to a house that was next door to the gray house. She was at the door for at least five minutes, and Aaron and the rest of them could only speculate on what she was doing.

Finally, she turned and walked back toward the cars. As she got close enough, she spoke. "You're right, James. The gray house is his, and he's not home. The neighbor says he's away at his summer home, somewhere on the coast, she thought. First thing we better do is to ditch some of these cars. It looks like some kind of caravan or something."

Aaron suggested they go eat and wait for darkness to fall. They could leave two vehicles at the restaurant and bring just one when they returned.

The restaurant was old, but clean. They all had burgers and fries, with the exception of Emily, who had a salad. They talked in a low tone and made plans. They decided that it was unlikely there would be an alarm. They also decided that if the police caught them breaking into the house, they would simply show the police all of their information and explain that they were just trying to verify it.

After they were finished eating, they took Herb's car and drove to a large food/variety store. Aaron went in and bought three pairs of gloves, two penlight-type flashlights, and a cheap pair of walkie-talkies. It wasn't quite dark, so they did more planning as they drove. If need be, they would break a window to gain entry, but first they would try all of the doors and windows. James insisted he could get in most places with just a credit card. The others were somewhat skeptical.

They parked the car on an adjacent road with a good view of the house. All roads led to the house. It was decided that Aaron, Herb, and James would be the ones to enter the house. Aaron would have rather had Emily along, but thought it best to keep an eye on James. Emily grabbed Aaron as he was getting out of the car and kissed him. As their lips parted, she whispered to him, "Be careful. I want you back in one piece."

"Don't worry. When faced with danger, I can run really fast. Keep that walkie-talkie on, and if there is even the smallest sign of a problem, let us know."

It was dark, and there wasn't much of a moon. The area was far enough out of town that there were no streetlights. They stayed together and moved to the rear of the house. Aaron and Herb stayed hidden behind some lilac bushes while James tried his credit card trick on the back door. In a minute, he returned and sounded out of breath. He started whispering, but was loud for a whisper. Aaron threw his hand over James' mouth and spoke in a low whisper. "You're too loud. Settle down and tell us what happened."

This time, James spoke more quietly. "When I went to slide my card into the latch, the door opened. It wasn't even locked. That makes me really nervous. Do you think there is someone there?"

Herb answered, "That's highly unlikely. There are no lights, no car, and the neighbor said he was gone. I say we go in."

"Me, too," whispered Aaron. James was somewhat reluctant, but finally gave the thumbs-up. Aaron suggested they stay together, and moved toward the back door with James in the middle and Herb bringing up the rear. No sooner had they entered the house than they heard the crackle of the walkie-talkie.

"What is it?"

Emily's reply sent them to the floor.

"There are two people standing right in front of the house, looking at it," she said. About a minute later, Emily gave the all clear.

Aaron could hear James breathing, and knew he was scared to death. He thought about Herb's gas can story, and about James wetting his britches. He started to laugh, and Herb put his hand on Aaron's shoulder and told him to shut up. Aaron turned and whispered, "I'm sorry. I just thought of something funny."

They waited until about midnight before they started to look around. Aaron turned on the little penlight and called Emily. "Can you see the light?"

"Just barely. If I weren't looking, I'd never notice it."

"Okay. Keep watching. We're going to start looking around."

Herb had the other light, and as he turned it on he gasped, "Oh my God! Somebody beat us to it." The place was in a shambles. The drawers were all pulled out and were lying on the floor. The cushions were pulled from the sofa and strewn about the room. Whoever was here hadn't cared about the way they had left the place. The whole house was a mess. They went from the den to the bedroom, and even checked out the bathroom.

Aaron motioned, and James and Herb followed him through the hall that led to a staircase to the basement. As they reached the bottom of the stairs, Aaron turned and whispered. "I don't know who the hell was here, but they were thorough. Doesn't look like something a burglar would do. It looks like they left all the good stuff.

If I had to guess, I'd say they might have been looking for the same kind of information we are."

"What now?" Herb asked. "Are we wasting our time?"

"Sure looks like it to me. I'd guess whoever did this knew what they were doing." As Herb and Aaron discussed the situation, James noticed something that didn't look quite right. He asked Herb for his flashlight and went to a side of the room that was a large brick wall, with a fireplace at one end. There was a mantle on the fireplace, and several other protrusions on the brick wall. James went to the far upper corner, away from the fireplace, and reached for a small hook that was embedded into the mortar. He pulled and nothing happened. He tried again with the same results.

"What do think you're doing?" Herb asked.

James explained. "Look at the edges of these four bricks. There's a hairline crack around them. You can't notice it from here, but stand over where we were, and take a look."

Aaron and Herb walked to the other end of the room and through the dim light they saw what James had noticed. There were four bricks, and the line jogged around them just like James had said. In the middle of the irregular set of bricks was the hook-like object, which protruded about three inches out from the wall. There were other assorted things stuck in the wall, and had it not been for the light hitting the wall just right, no one would ever have noticed the bricks. Aaron handed Herb his light and asked him to hold it on the bricks. As Herb and James held their lights on the spot, Aaron began to move the handle back and forth. The handle seemed stuck, so Aaron applied more pressure. Suddenly, the handle moved, and the bricks moved easily back and forth. Aaron pulled, and the bricks came out of the wall in one piece. They weren't actually bricks, but rather an odd shaped drawer faced with thin brick fronts. The drawer was about six inches deep, and it was completely empty. Aaron sat the drawer on the floor and took his light from James. He shined the light into the cavity behind the drawer. There, about a foot into the wall, was a dusty briefcase. The handle faced the opening as it lay on a flat surface. Aaron didn't reach for the handle, but looked for something to hook it with. James found a metal coat hanger in a closet and handed it to Aaron. Aaron took it apart and bent it out straight, leaving the hook on the end. He then reached the hooked end into the cavity, holding the coat hanger in one hand and the light in the other. He hooked the handle of the briefcase and moved to the side of the cavity. He asked James and Herb to stand to the side.

Herb was getting impatient. "What the hell are you doing? Do you think it's booby-trapped?"

"You never know. I saw this in a movie." Aaron pulled, and the case came out from the cavity. It was dusty, and looked like an antique. Aaron lowered the case to the floor and undid the latch. Inside was what appeared to be some sort of a manual, written in a language that none of the three could recognize. Aaron looked into the cavity and saw something that made his heart skip a beat. There, in the dim light, was a pistol in a holster. Aaron left the pistol in the cavity and very carefully replaced the empty briefcase back into the cavity. He then took the drawer and tried to insert it back into the wall. He had a hard time lining the drawer up to the cavity. Once it lined up, it slid smoothly into place. They took one last look around and then made their way out of the house and back to the car.

Emily acted like a kid at Christmas. "What did you find?"

Aaron thought for a second. "We have no idea. It was hidden in a brick wall, and it isn't in English."

"How did you find it if it was hidden in a wall?" asked Rachael.

James had a quick comeback. "I saw what looked like a pencil line on a brick wall. When we messed around with it, it opened. There was a pistol in the wall, but we left it there. I'll bet old Finley pulls that pistol out when he gets home and goes looking for the people that tore his house to pieces."

Emily had a kind of a weird look on her face. "What do you mean the people that tore his house to pieces?"

"Oh yeah," Aaron said. "I almost forgot. Somebody got there before us and tore the shit out of that house. I can only guess who it might have been. One thing though, I bet they wish they had this book."

Herb drove back to the restaurant and parked next to Aaron's pickup. "Well, so far, so good. I'm sorry I tried to get out of this the easy way," James said. "I'm really glad you guys talked me out of it. Who's gonna keep the book?"

Emily spoke up. "I think Aaron should keep it, seeing as how he's the least involved." Everyone agreed. Aaron would keep the book and try to find out what it was.

They decided to meet again in three weeks, instead of the usual month. Herb, James, and Rachael would try to get a lead on who else might be so interested in the professor. Rachael asked Aaron and Emily to stay the night, but they declined and drove all the way back to the motel in Bend.

It was after noon when Aaron finally awakened. Emily was gone, and he assumed she had gone for a run. He slowly got up from the bed and went into the bathroom. He looked in the mirror and had to laugh. They had both gone to sleep fully clothed, lying on top of the covers, and he now looked like a transient. His hair was messed up, and his shirt was wrinkled.

He was about to jump into the shower when he remembered the book. He put his shoes on and picked his keys up off the nightstand. The book was there, lying in plain sight in the front of the truck. He took the book back into the room and thumbed through it, page by page. He had no idea what he was looking at. About halfway through the book, there were notes written alongside the script, some in English, and some in the other language.

Aaron's heart started to beat faster as he read. Then, there in the upper corner of one of the pages, was the name Marty Smith with a line under it. Under the line was the word "governor." Aaron sat, staring at the name and the word "governor." This, he thought, was the reason there was so much interest in the professor's house. It had something to do with the governor. As he went on through the book, there were other names and other notations. He recognized the name of the current Democratic U.S. senator, and under his name it read "senate or house." There was a knock on the door, and Aaron quickly put the book under the bed. As he opened the door, Emily burst in and threw her arms around him. "I feel good. You should have come with me."

"Wait 'til you see what I've found." He pulled the book from under the bed and found the page with the governor's name on it. "Take a look at this, Emily. I think we've hit the jackpot."

"Oh my God, I think you're right."

After they both took showers, they went to brunch in the motel restaurant. Aaron overdid the buffet, while Emily had a croissant and a cup of black coffee. He admired her for her self-control when it came to eating. He thought that he'd probably have to start watching what he ate as soon as he retired from the woods, in 50 years or so.

They were in a booth and talked in a low voice. "Emily, what do think we should do? Any ideas?"

"You once told me that Mary was a whiz at foreign languages. Call her first to see if she knows someone who can translate this book."

"You know something, Emily? I love you. Will you marry me?"

Emily stared at Aaron and didn't say a word for several seconds. "You're serious, aren't you?"

"Not the answer I was looking for. Did that catch you a little off-guard?"

"Yes...I mean, yes, it caught me a little off-guard. I mean...yes, it caught me off-guard, and yes, I'll marry you. Today, if you'd like."

The water glass in Aaron's hand began to shake a little, and he had the sudden urge to pee. He excused himself and went to the bathroom. All he could think about was where the hell that had come from, and now what should he do? He was more composed when he got back to the table. "So what's good for you?"

Emily thought for a second. "How about Christmas? That way we'll only have half a school year left. You'll be done in May, and I'll only have student teaching left to do. I can do that in LaGrande."

"My dad is gonna be shocked, but Mary and Mom are gonna be ecstatic. I can already hear the planning episodes."

"You think your folks will be shocked? My dad thinks I'll never get married. Aaron, I hope nothing happens to mess up our lives."

"I wouldn't say we are out of the woods yet, but I've got a hunch we're on the right track."

As soon as they got back to the motel, it was time to check out. They packed up their stuff and loaded up their vehicles. They stood and kissed before they left the room. Aaron wanted to stay, but Emily didn't want to pay for another night's lodging. Aaron kissed her through the Explorer window and mouthed, "I Love You" as she backed out of the parking lot.

Chapter XII

As Emily started over the Santiam Pass, she called home on her cell phone. Her dad sounded different. "Hi, Honey. Are you in some kind of trouble?"

"Why are you asking me that?"

"Oh, I don't know; just thought I'd ask. There were a couple of state troopers here that wanted to talk to you. They said you might have information they need."

"I don't know what they think I know. What else did they say?"

"They said they'd be back."

"The phone is breaking up, Dad. I'll talk to you when I get home." Emily's heart raced, and her mind was spinning as she drove over the pass. All she could think about was how stupid she'd been to get involved with FF. Then she rationalized that if she hadn't gotten involved with them, she would never have met the love of her life.

When she reached the summit, she tried calling Aaron. The connection wasn't really clear, but the message got through to Aaron loud and clear. Aaron listened and then tried to calm her down. "Don't worry; just tell them you don't know anything. I'll call Herb and tell them what's going on. I'll call you when I get home. Are you still gonna be my wife in December?"

"Oh Aaron, I hope so."

When Aaron turned north at Mount Vernon, he checked the signal on his phone. It was working, so he pulled over and dialed Herb. Herb answered, and Aaron was relieved. "Hey Herb, this is Aaron. We've got trouble."

" Oh, hi, Mom! Can I call you back in a little while? I have company. Okay. I'll call you later." The receiver clicked, and Aaron sat there on the side of the road, totally stunned.

He thought for a minute, and then called home. Nancy answered, and he asked to speak to Mary. "Anything wrong, Aaron?"

" No, Mom. Why??"

"Nothing. You don't normally call us on your way home. Will you be here for dinner?"

"No, I won't be home until 8:30 or 9:00; I just need to ask Mary for a favor."

"Just a minute. I'll get her."

When Mary came on the line she asked, "How was the trip?"

"It's still in progress. I'm only in Mount Vernon. Will you do me a big favor, and meet me at the mill office at 8:30?"

"Sure, what's up?"

"Just meet me there."

Mary was already at the office when Aaron pulled in at 8:20. He grabbed the book and went into the office. "Hi, Mary. Thanks for coming."

"You're more than welcome. Now what's up?" Aaron started telling Mary the story as he dialed Herb.

Herb answered, and before Aaron could even say hello, he started talking. "I'm sorry. I can't hear you very well. Call me back on my cell. Something must be wrong with my phone." Aaron dug into his wallet and found Herb's cell number and called him back.

"Sorry, partner. I think the land lines are bugged."

"What's going on there? Who came to see you?"

"Two detectives from the state. They said they had reliable information that I was involved in the burning of the Oak Ridge ranger station. I pretty much told them to piss off. Has anyone paid a visit to you yet?"

"No, but they were at Emily's parents' place this afternoon. Do you think they know what we're doing?"

"I don't know, Aaron, but whatever you do, don't let anyone get that book."

"Don't worry; no one will get it."

"What was that all about?"

"Mary, help me copy this entire book while I tell you one incredible story." Aaron told her the whole story of how they had obtained the book. Mary retrieved the pages from the copier and put them facedown in a neat stack.

When Aaron finished the story, Mary hit the button on the copier and turned to Aaron. "What in the fuck do you think you're doing? This is like some kind of a James Bond movie." Aaron looked at her and started laughing.

Mary found nothing humorous about the situation. "What on earth do you find so funny?"

"I've never heard you say 'fuck; before; I wasn't sure you'd ever heard the word before."

Mary's face turned red. "I'm sorry, but you're scaring me."

"Mary, if I said that I wasn't scared, I'd be lying. I'm scared as hell, but not for us. I'm scared that Emily might be in danger. You know, she is my fiancée."

"What?"

"You heard me. I asked her to marry me, and she said yes. I just hope nothing happens to mess it up."

"Congratulations! When's the wedding?"

"Some time around Christmas."

They finished copying the book just before eleven. As soon as they were done, Aaron called Emily on her cell phone. "Hi," Emily said. "I wondered when you were gonna call. Did you get the gist of what went on over here?"

"Yeah, the same thing went on at Herb's."

"I know. I talked to Rachael, and no one's contacted James or her yet. What do you think is going on?"

"I don't really know, but don't give anyone any information, especially about the book."

"You still want to marry me?"

"I do. I guess I better practice those two words. I love you. Talk to you tomorrow."

"Love you too, bye."

After he hung up, Aaron and Mary talked about what the book might be. Mary was sure that the writing was Russian. They thought about where to hide the original. They decided to bury it at the edge of the log yard. Aaron took the keys to the maintenance shed and then went to the lunchroom. He took a roll of foil and wrapped the book in several layers. He then wrapped it in layers of plastic wrap. He placed the wrapped book into a small box he had found in the supply room. He then wrapped the box in plastic wrap and foil. After it was ready to bury, they went to the maintenance shed and took a shovel and tape measure. Mary made a map of the spot. Aaron took a tape measure and took measurements from a tall pine tree that was at the north end of the log yard. He measured 10 feet from the tree in a direct line with the sign at the mill entrance. Aaron started digging. He dug a hole about two feet deep and placed the box in the bottom of it. He then covered the box with a piece of bark and filled the hole in with dirt. "We don't need a map, Mary. Just remember that the hole is exactly 10 feet from this tree in a direct line with the sign. If anything happens to me, tell Dad where the book is."

"Nothing's gonna happen to you, okay?"

"Okay, but just remember what I said. Ten feet from the tree."

Aaron followed Mary back to the ranch, and they hid the copy in an old feed can in the barn. They talked until 1 A.M. about what to do with the copy. Mary suggested that she take a page or two to a professor who taught at the university in town. Aaron asked about the professor, and Mary confided that she did not know him very well. He was new, and Mary had taken a German night class from him. Aaron didn't want anyone to have access to that much of the book just yet. He suggested that they copy the title and see if they could at least get a general idea of what it was about. Mary thought that sounded like a good idea. She would do it first thing Monday, if Nancy would give her a little time off.

Mary left the office at 10:30 and made it to Eastern just before eleven. She went to the office and asked the receptionist if Doctor Howley would be able to see her for a moment. The lady picked up the intercom and called Dr. Howley's classroom. "Doctor, there is a young lady who would like a moment of your time. She says that she took a German class from you last term."

There was a short pause, then a response. "What kind of a grade did she get?" The receptionist looked at Mary, and Mary mouthed the letter 'A'.

"She got an 'A', Doctor."

"Have her come down to Room 107."

"Thank you." Mary found the professor sitting behind his desk, looking over some papers.

"Well, hello, Mary. Why didn't you just have her tell me who you were?"

"I really didn't think you'd remember me."

"I gave out three 'A's' to your German class; I always remember the 'A's'. What can I do for you?" Mary took the piece of paper from her purse and handed it to the doctor. The first word had what looked like two 'O's', a weird-looking 'W', a small 'E', and then "CTBeHHbIH." The second word looked somewhat the same.

The doctor looked at the paper, and then looked back at Mary. "Can I ask you why you want to know what this says?" Mary was ready for the question.

"I saw it on the computer but couldn't find the meaning. Is it Russian?"

"Very Russian. It says *Social Manipulation*. Don't you find that a little strange?"

"Very strange. I wonder what it means?"

"I think that would depend on where it was used." Mary asked the doctor if he knew of anyone who did translations. He suggested a high school student by the name of Dave Hill. He told Mary that Dave was more familiar with the Russian language than he was. Mary knew Dave.

She thanked the doctor and started toward the door. "Oh, Mary," the doctor said. "If you need anything else, don't hesitate to ask."

Mary drove straight to the mill office and grabbed the phone book. She found an E.P. Hill listed and called the number. There was a recording, so she left a message asking Dave, if this was the correct household, to please call her. She left both the mill and home numbers.

Just before quitting time, Mary received a call. "Hello?"

"Hi, Mary. This is Dave. Did you call me?"

"Yes. How's your summer going?"

"Except for working at The Burger Hut, pretty well. How's yours?"

"Pretty good, except for working at the mill. Dave, I may have a favor to ask you. Dr. Howley says you're a whiz when it comes to Russian."

"I should be. My mom was born in Moscow. I spoke fairly fluent Russian by the time I was five."

"If I had a book that was written in Russian, would you be able to translate it into English?"

"I could, but there probably wouldn't be a need to. Most books are written in more than one language."

"I don't think the one I'm thinking of is. As a matter of fact, I'm sure it isn't. I don't know that I could pay you a lot, but I think you might find it really interesting."

"I work about six hours a day, and the rest of the time I play X-Box or watch TV. I'm pretty sure I could find some extra time."

"Good. Let me talk to Aaron, and I'll call you back. Will you be around this evening?"

"I should be. Give me a call."

"Thanks, Dave. Bye."

"Goodbye, Mary."

Aaron didn't get home until after six. Mary was waiting on the front porch. Aaron could sense good news. "What did you find out?"

"Oh, not much. Just what the title to your book, or manual, is."

"So what is it?"

"*Social Manipulation.* Isn't that creepy?"

"Worse than creepy, Mary; it's sinister. These dirty sons of bitches have been influencing us for almost 50 years, and they've done one hell of a job. I can't wait to find out what's in the book."

Mary told Aaron about Dave, and Aaron thought that it was a great idea. They decided that Mary would arrange a meeting with Dave. They would both get a feeling as to whether or not Dave could be trusted.

Mary called Dave and arranged a meeting. They met Dave at the high school parking lot just before dark. Aaron started the conversation. "Hi Dave, I'm Aaron, Mary's brother. Mary tells me you know Russian inside and out."

Dave looked at Aaron, and then at Mary. "What's going on?"

"We need to have a book translated from Russian into English, and we need it done with complete secrecy. The only thing I can tell you about the book is that we found it, and most likely shouldn't have it. If you do this, I'll give you $200, but we need to do it together, page by page. I know that doesn't sound like a lot of cash, but it's about the best I can do." Dave walked around Aaron's truck and stared off into space.

He finally answered. "I'll do it for nothing if it's worth doing. If it's boring, I'll just forget about it." Mary walked over to Dave and put her hand on his shoulder.

"That's great. I've got a hunch you'll find this particular book very interesting." Dave rarely had to work after five, which fit Aaron and Mary's schedule. They made plans to work on the translating at seven each evening. They would meet at the mill office and work until 10 P.M. each evening. Dave would translate the words, and Mary would type the translations. Aaron would make sure everything stayed in perfect order.

There was a major thunderstorm Tuesday afternoon, and it was still raining when they met at the mill. Mary had everything arranged. She had dug an old Royal typewriter out of the supply room and scrounged up a ribbon. She was an excellent typist and hoped that Dave wouldn't take a long time translating the words. They didn't want to use one of the office computers for fear of someone finding the translation. Aaron pulled A.J.'s chair out of his office for Dave to sit in.

He handed Dave the copied book and sat on the office counter. Mary was sitting at Nancy's desk, waiting for Dave to start talking. Dave read the book but didn't say a word. As he read, Mary and Aaron looked at each other, not really knowing what to say. Dave finally looked up and spoke. "Do you guys have any idea what you have here?"

Aaron jumped off the counter and walked over to Dave. "We have some idea, but why don't you enlighten us?"

"This is a Russian manual explaining how to destroy the U.S.A. It's very descriptive. Listen to this." He took a deep breath and began to read.

"You, as American citizens, have the power to destroy the social infrastructure of America. You must not fail in your duties. In each institution, you will have the opportunity to mold thousands of young minds. The following information in this manual will explain in detail your mission. Do not deviate from this manual, as the research completed by the social architects is conclusive and accurate." Dave stood up and looked at Aaron.

"Where did you get this? Do you know where the original is?"

"Look, Dave, we just need you to translate this thing, not decipher its meaning. Will you or won't you do that?"

"Hell yes, I'll do it. This is, by far, the most interesting Russian book I've ever seen, and I've seen a lot of them."

Dave read, and Mary typed. Aaron took each page and put it in order. Mary was amazed at how fast Dave translated the words. His delivery was so smooth that you would have thought he was reading English. By 10 P.M., they were already on page 20. Aaron had counted the copied pages and found that there were 207 in total.

"We better hang it up," Aaron said as he put the typed pages into a folder.

"We're just getting to the good stuff," Dave said. "Why don't we go a little longer?" Mary looked at Aaron and nodded.

Dave started reading. What he read shocked Mary, and she stopped typing. "Choose the most intelligent students. They are the weapons we seek. They must be given every advantage to gain entry into the field you choose for them. Educators will be your first priority. You must make sure that the educators you choose will teach as you have taught them. First, you must gain their confidence, making them feel that they could change the world if just given a chance. You and you alone must give them that chance." He looked up. "Wow," he said. "This is pretty heavy stuff. Do you think this is real, or just fiction?"

Aaron didn't hesitate. "It's real, Dave. As real as death."

They finished at 10:45 and vowed to meet back at 7 P.M. the next evening. After Dave left, Mary and Aaron shredded the finished pages and put the remaining pages back into a plastic zip-loc bag and placed it in the feed can. Aaron called Emily, using his cell phone. He told her about the translation and some of the information in the book.

Emily was amazed at the things Aaron told her. "What are we going to do when the translation is finished? Do you think we should give it to the police?"

"I really don't know. We'll cross that bridge when we come to it. Have you had any visitors?"

"Yes, they were here today. All they wanted was information about Herb. They didn't say he'd done anything. They just asked some general questions."

"What did they ask you?"

"Basically, they just wanted to know if I knew him. They did ask one question that I thought was a little funny."

"What was that?"

"They asked me whether or not I knew if Herb knew Professor Jonathon Finley."

"What did you tell them?"

"I told them I had no idea."

"Good. I'll talk to you tomorrow. Love you."

"Love you too, bye."

By Friday at ten, the book was over half finished. Dave was doing a remarkable job. There were parts that weren't a 100 percent true translation, but Dave was such a student of Russian that he easily filled in the blanks. The content was mainly disregarded, with the urgency to finish the project the priority. There were parts that stopped them, and small discussions occurred.

There was a part about the formation and growth of unions that Aaron reflected on. That part read: "Unions must be protected and encouraged to grow. They must be three times more powerful than the business or government agency that they are affiliated with. Through organized bargaining, economic destruction of both private enterprise and government can be achieved."

"Wow," Aaron said. "Now I know why Dad was so adamant about not letting the union in here."

Dave took that as a bit of a challenge. "You know, unions do a lot of good. Without them, people would be taken advantage of. My mom thinks the teachers' union is the only reason she gets paid for what she does. She says that without the union, they'd be screwed."

Mary interjected, "I think it's all about common ground. There should be give and take both ways. I do think you should be paid on merit though, and not because you become a protected fixture somewhere."

"Let's finish this thing and then discuss it."

"You're right, Aaron," Dave said. "Let's do it."

The translation was completed two weeks from the day they started. The copy that was in Russian had been destroyed, and the English version was complete. It was 9 P.M. Tuesday evening, and the office was still warm from the day's heat.

Aaron went out to his truck and came back in with a cooler. He pulled out three ice-cold near beers and gave them each one. Dave had the twist top off of his before Aaron had a chance to propose the toast. "Hang on a second. I want to take this opportunity to thank the incredible David Hill. We have no idea if what you translated is or is not correct, but we think it is."

"Thank you for your vote of confidence, Aaron. I also have no idea if it's correct." They toasted and drank their beer. Aaron took the empty bottles and put them back into the cooler. Aaron didn't really know how to properly thank Dave, or how to repay him for the tremendous effort. Aaron put the completed English version into the zip-loc bag and placed it into the old feed can. He would hide it in the same place the copy had been hidden. Only Mary and he would know its whereabouts. He was ready to leave, and Mary informed him that Dave was going to take her home. As Aaron pulled away from the mill, his thoughts were of Emily and of what strange circumstances sometimes bring people together.

Chapter XIII

As Jonathon Finley pulled into the driveway of his home, he took note of what a good job the boy had done on his yard. He was actually in good spirits until he pushed the key into his back door, only to find it already opened. Professor Finley had rarely known the sensation of being afraid.

The sight of his cluttered belongings sent chills down his spine. He moved as quickly as he could to the stairs leading to the basement and carefully made his way to the bottom. He moved to the far end of the brick wall and slowly turned the hook that latched the brick-faced drawer. The drawer easily slid open, and the old professor breathed a sigh of relief when he found his satchel still in the cavity. As he raised the drawer to replace it, he noticed that the dust under the old leather briefcase had been disturbed. He took it from the wall and, slowly opening it, he cried out in Russian, "Het, Het, Het."

He took the revolver from the cavity and replaced the briefcase, not bothering to put the cover back in its place. He climbed the stairs and walked to the kitchen. He picked up the receiver and dialed from memory. "Governor Smith's office. May I help you?" Professor Finley hung up the phone. He put the revolver in his coat pocket and walked out to his car. He was in a state of mind that he had never before experienced. He knew that the government had his manual and that surely it would be just a matter of time before the whole world knew of the project. Over and over in his mind he cursed the bumbling Marty Smith. How had that idiot been able to recognize what had happened to him? How had any of them had the courage to contest the person that was responsible for all of their glorious power?

He parked in the employee lot, took the pistol from his pocket and placed it in the glove box. He walked briskly toward the capitol building and through the large entrance. The receptionist knew the professor and greeted him as he stood before her desk. "How may I help you, Professor Finley?"

"I'm here to see Governor Smith. Will you please tell him I'm here?"

"I'm sorry, but the governor isn't receiving visitors today. If you'd like, I'll give him a message."

"Yes, do that, and do that right now. Tell him that if he doesn't see me right now, I'll tell the entire world of his extramarital affair with a certain state senator." The receptionist left her desk and went around the corner toward the governor's office.

She returned a few seconds later and motioned the professor to follow her. She led him to the entrance to the governor's office and then left. Marty Smith stood behind his desk, and the professor could see that he was irate. "Nice, Jonathon. When did you add blackmail to your repertoire?"

"It's not blackmail, you little bastard. It is a matter of life and death. My physical existence and your political existence."

"What the hell are you talking about?"

"I'm only going to ask you this one time. Where is the book you took from my house?"

"I have no idea what book you're talking about. Have you gone mad?"

"I told you I'd only ask you one time. Now suffer the consequences." The professor headed toward the door, and Marty was quick enough to get between it and the professor.

"Wait a minute. Sit down, and I'll tell you what took place." The professor took a seat in one of the two chairs on the visiting side of the governor's desk, and Marty sat in the other.

"What I'm going to tell you will no doubt infuriate you, but I think it was done with good reason. Hear me out, and then defend yourself."

The governor told the professor everything, from the secret meetings to the search of his house. He was convincing when he told the professor that nothing had been taken from his house. He went into great detail about the research that had been done on his immigrant parents, and his brother. The story actually seemed intriguing to the professor.

The more of the story the governor told, the more interested the professor seemed to be. Finally, the governor came to a sudden stop. "Do you find this amusing, Professor?"

"Why, as a matter of fact, yes. I've spent a number of years helping you destroy your beloved state here, and it is only now that I see just how effective I've been."

"Who the hell are you?"

"I'm who I seem to be. An American-born U.S. citizen, who has rarely, if ever, broke the law. A college professor who has changed the minds of so many, and helped mold the minds of so many more. A knight in shining armor to so many whose political careers I've enhanced. If the book that was taken from my basement isn't recovered, I'll be the biggest traitor since Ethel and Julius Rosenberg. And you, my good man, will be my accomplice."

Marty sat still and tried to decipher what he had just been told. He didn't give a damn about what the professor had done or who he was. He cared about himself, and about his own political career. "So what's in the book, Jonathon? Is it some kind of a journal?"

"No, it's a manual, with footnotes and names scribbled in the margins. Names like Marty Smith."

"I don't know who took your book, but I'll get it back, that I can promise." Professor Finley told the governor to call as soon as he had any information. He warned against any sort of retaliatory response, and the governor assured him that there would be none.

Before the professor had cleared the building, the governor was on the phone to his special agents. The same people who had ransacked the professor's house would now set out to find the book that the professor so desperately needed. The small band of officers immediately started the process of recovering the manual.

Chapter XIV

This time, the meeting was in Prineville, and Aaron was there just after eight on Friday night. Just before nine, Emily's Explorer pulled into the motel. She pulled her phone out and called Aaron. "What room are you in?"

When she got to Room 228, she had an overnight bag in her hand, which Aaron took from her and set on a chair. Before he had a chance to say hello, she had his shirt unbuttoned and half off. She had her hand down the front of his Levi's, and he started laughing.

"What's so funny?"

"Nothing. I was just wondering if you had missed me?" She looked into his eyes and he knew the answer. He pulled her tee shirt over her head and then reached around and undid her bra. His hands moved gently over her breast as he kissed her neck. He then moved his tongue from one nipple to the other, and then over her bare stomach to her navel. She was unbuttoning the last of his Levi buttons when her cell phone rang.

"Don't answer it"

"I have to. They know we're here because our cars are out front." She grabbed the phone. "Hello?"

"Miss Alcott, are you busy?"

"No, Herb. We're in Room 228. Come on up."

James had ridden with Herb and Rachael. Herb, James, and Aaron sat at the small table, and Emily and Rachael sat on the bed. Aaron reached into his overnight bag and pulled some papers out and laid them on the table. He then took the old feed can out and opened the plastic lid.

He took the translated manual out and laid it on the table. He took some of the papers from the table and gave one to Emily and one to Herb. "Protect these. This is a map of where this copy is hidden and also a map of where the original is buried. I thought we should make sure that this information is never lost. If something would happen to Mary and me, you guys would know where to find the books. Well, here it is, the translated copy of the manual that is destroying the United States. It's all here, every dirty little trick they used to get the ball rolling in the wrong direction. It is very specific in some regards, and very general in others. I've noted some interesting parts for your reading enjoyment. The pages are all numbered, and I've handwritten in all the notes that were in the original."

Herb sat and paged through the stack of papers. He read aloud when he came to a part that he considered important. The five of them sat there for an hour. Herb read while the others listened. "It is important that you find individuals who can be molded into what we need. They must be of high intelligence and have little common sense. Take 10 from each class and concentrate on them. If they are not conforming, release them and try to replace them with others who will conform. It is important to give instruction to all students that you are in contact with, but the special 10 are to be your prime objectives." Herb stacked the papers neatly and sat back in his chair.

"Wow, I guess I was one of the special 10. Makes me feel all warm and fuzzy inside. That rotten son of a bitch was so good at what he did. I often wondered why I was doing the things I did. Now I know."

"So," said Herb. "What do we do with it? This is probably the most important document that exists, except for the Bible. I also think it might be the most sought after when the professor finds it missing."

"Who," Rachael questioned, "will seek its return? One aging professor?"

Herb answered. "I don't know. Maybe the KGB will be after it. Is there anything in here about the media being so liberal?"

Aaron thought for a moment. "Look toward the very back of the book, like the last 20 pages or so." Herb paged through the last part of the book until he found what he thought to be interesting.

"Listen to this. This is great." He paused to ensure that he had their attention, and then read.

"Promote a sense of self-righteousness in all of your journalism students, for they will be the messengers to the masses. They must be fighters who will stop at nothing in order to destroy the conservative establishment. Instill a mindset that will promote liberalism in any message they give. They are the best weapons we have." Herb set the book down and stood up. The others were staring at him, as if he had made the statement up.

Finally, Aaron spoke. "No, that's real. I remember that part." Herb started laughing, and then they all laughed. Herb leaned over the table and extended his hand to Aaron.

"You did an incredible job on this thing."

"Thanks, but I'll give most of the credit to Dave and Mary. Without them, we'd still be looking at a bunch of funny-looking letters. Bottom line though, it's done, and we need to figure out what to do with it."

They all went across the street to a small restaurant for a late dinner. The conversation drifted to the people who had been after Herb and Emily. James told the group that he'd found some interesting information about the professor and his connections to various state and federal legislators.

A reliable source had told him that Professor Finley and Governor Smith were more than casual acquaintances. It was the father of one of James' closet friends, and he'd been a classmate of Marty Smith's. It seemed that there were many private meetings between the two during Marty's last year of graduate school. Two years later, Marty was a state representative. He proposed more liberal legislation than other members of the state legislature.

"My friend's dad said that it was like the professor had a microphone built into Marty, and as he spoke, the words came out of Marty's mouth, but the messages were the professor's."

"That's not really big news, buddy." Herb hesitated. "The message that Finley pushed came out of a thousand mouths, mine included."

After they finished dinner, they went back to Aaron's room to continue the meeting. The major topic was the people that were questioning Herb and Emily. Neither James nor Rachael had been approached.

They decided to distance themselves from Professor Finley, and ask no more questions. They would take a "wait and see" approach. Instead of trying to find out how the professor and the government were involved, they would wait until the government made the next move.

The next meeting of the group would be the first Saturday in September. It was understood that by the next meeting, each member of the group would have a valid idea of what to do with the book. They would weigh each idea and choose what the group, as a whole, deemed the best. Until the next meeting, they would communicate by cell phone only. Before they broke the meeting up, James asked a troubling question. "How much do you think the book's worth?"

Social Manipulation

Aaron and Herb both approached James. Aaron spoke first. "Don't even think about it. If you say one word to another soul, I'll break your neck."

"And I'll help him," added Herb.

James assured them that he had no intention of trying to profit from the book. He said that it was a question borne only out of curiosity.

Emily and Aaron spent a wonderful night wrapped in each other's arms. After the lovemaking was over, Emily hit Aaron with a proposition. "I know you are going to think I'm old-fashioned, and I know it sounds stupid and probably doesn't make any difference, but what would you say to celibacy until we get married?"

"I'd say that you're right. It does sound stupid. But if that would make you happy, then I'll try to control my raging hormones for a few months. Tell me it's not because you don't like it."

"I love it, and when we get married, we'll do it twice a day until we're 50."

They met at the little restaurant at nine and ate breakfast together. They knew they were in deep trouble and that to survive their ordeal, they might have to use the book as leverage. There was still the matter of the ranger station, and all of the other things that Herb, Rachael and James had done.

Aaron suggested that revealing what was in the book was the most important thing of all. Herb had pointed out that their lives were really the most important thing. His thinking was that all of the damage was done, and that the book would only serve to tarnish the liberal agenda. He was worried, and Aaron sensed it. "Look, Herb, if it means saving your ass from jail, I'd look at a deal. We just need to let people know what has taken place over the past 50 years. All of those fucking movie stars that think they're saving the world need to know what idiots they really are."

After the others left, Aaron and Emily went to the room and started making some preliminary wedding plans. It wasn't long before it was decided that Emily, Trish, Nancy, and Mary would actually do the planning. Aaron would lend any technical advice. They were checked out and ready to leave by noon. Aaron wanted to stay another night but had promised A.J. that he'd help Jed on Sunday, since things at work had been so busy. Aaron kissed Emily goodbye and watched the Explorer as it went out of sight. He threw his overnight bag in the back of his truck and headed for home.

Chapter XV

Emily never saw it coming. The lady flagging her down looked normal. She was just a poor woman in need of a hand. Emily stopped in front of the Jeep Cherokee and looked into her rearview mirror. The hood was up, and the lady stood by the driver's door. Emily got out and started walking toward the woman. As she approached, something seemed wrong. The woman's hands were large, and she could see an abnormal amount of hair on her arms. Emily looked up and down the highway for other cars, but there were none. She looked the person in the eye and knew it wasn't a woman in distress; it was a man. Emily turned and started running toward the Explorer. As she approached the driver's side door, a man appeared from the rear of her rig and grabbed her. She started screaming and kicking, but before she knew what hit her, the man that was dressed like a woman shoved a rag in her face, and she passed out.

When she regained consciousness, there was a piece of tape over her mouth and her hands and feet were bound with tape. There was a thin blanket over her, and when she opened her eyes she could see that she was in the back of a moving vehicle. She started to struggle against her restraints, but it was no use. As she lay there, she could tell that the road was getting much rougher, and that the vehicle was moving much more slowly. Finally, it came to a complete stop, and she heard the door open and close. She could see part of the rear door as it swung open. Someone took her by the arm and sat her upright. He then pulled her legs toward the opening, and cut the tape that bound her legs together. He pulled her by the arm, and she slid out of the back of the Jeep and stood up.

Without a word, he led her to a small cabin and opened the door. The guy was probably in his forties and seemed physically fit. He didn't look like the typical rapist type, and Emily breathed a sigh of relief when the guy finally spoke to her. "I'm not going to harm you in any way. Just don't do anything stupid, and you will be fine."

He sat her down in a chair and took a pair of leg irons from a bag. He put the irons on her legs, wrapping the chain around the chair brace. He then cut the tape on her hands and took a pair of handcuffs from the bag, cuffing her right hand to one of the legs of a nearby table. After he had finished, he took the bag and left the cabin.

Aaron came out of the little store in Long Creek with a piece of jerky and a Coke. He tried calling Emily, but the signal wasn't strong enough. As he pulled out and headed north on Highway 395, he thought about how wonderful it was going to be when they were married. He knew that it would be a little tough at first while they finished college, but he had been a good saver, and knew that they would be fine financially until he finished school and started working for the mill full time. He pictured the two of them taking Saturday trips up to the lake. He would have to get Emily her own horse. He thought about her celibacy idea and wasn't very fond of it. What was the difference? They had already done it. He finally came to the conclusion that it was a girl thing, and that A.J. was right. They did think differently than men did.

About five miles out of LaGrande, the signal on Aaron's phone was at full strength. As Emily's mailbox answered, he got a sick feeling in the bottom of his stomach. "Hi. Call me as soon as you get this message. I love you."

He drove by the mill and parked next to A.J.'s pickup. A.J. had worked Saturdays as long as Aaron could remember. He vowed that he was going to be a five-day-a-week guy, and a family man on the weekends. It had always been about work, and Aaron always felt he had missed a lot growing up. He loved his dad and felt blessed for the things they had done together; he'd just do more with his kids.

A.J. sat behind his desk, sketching on a piece of graph paper.

"What are you dreaming up now?" Aaron asked him.

"Oh, hi. How was your visit this time?"

"Good, Dad. Has Mary told you about Emily and my plans?"

"No, but I think she might have told Nancy something. What's going on?"

"Not much. We're getting married in December, that's all."

"Ordinarily, I'd say you were crazy, but seeing as how it's Emily, I think you're a lucky kid. Ex-tree-huggers make great wives."

"She was never a tree-hugger; she just got caught up in the "save the world" thing, but believe me, that's changed. Boy, how that's changed."

Aaron drove to the ranch and tried calling Emily on the way. He figured that she had forgotten to turn the phone on, and he would just wait for her to call him. Mary was out in the barn, brushing her horse. "Did you go for a ride?"

"Yeah, Nancy and I rode all the way over to Brush Creek and back. I'm going to be sore tomorrow. I told Nancy about you and Emily. I hope that's all right?"

"That's fine. What did she think?"

"That you two would be fine, and that planning a wedding was going to be fun. Have you decided where you are going to get married?"

"Emily wants to do it over here, but she has to get that approved by her parents first."

"Well, wherever you decide to have it, Nancy and I would like to help."

"Don't worry; Emily has already penciled you guys in."

Aaron walked to the house and picked up the phone in the den. He dialed Emily's phone again, and again she did not answer. He called her parents' number, and Don answered the phone. "Hi, Don. This is Aaron. Is Emily there?"

"No. What time did she leave?"

"About eleven. She should be there any time. Would you have her call me?"

"As soon as I see her."

"Thanks, Don. Bye."

"Goodbye, Aaron." Aaron knew that Emily should have been home over an hour ago, but he didn't share that with Don. He knew that something was wrong.

He called Herb, and Rachael answered. "Hey, Rachael, have you guys heard from Emily?"

"No, why?"

"She should have been home an hour ago, and she's not answering her phone. That's not like her."

"No, it's not. She always has that phone by her. Do you think something happened to her?"

"Jesus, Rachael, I sure hope not. Is Herb there?"

"He's up in the attic, hiding the map you gave him. He's a bit paranoid."

"Maybe he should be. If you hear from her, have her call me. Bye, Rachael."

"Bye, Aaron."

Emily asked her abductor if she could use the bathroom. He had a procedure ready for the occasion. He undid the leg irons, and Emily watched as he placed the key back into his pants pocket. He undid the table leg side of the handcuffs and led her to the tiny bathroom. There was a pipe that extended from the floor to the roof, and he clasped the cuff around it. He left and shut the door. Emily examined the room and took special note of the small window. The door had a small lock just within her reach, and she latched the door. She knew the guy could break the door down with a single shove, but she felt better that it was locked. When she finished, she unlatched and opened the door. The guy wasn't in sight, and she thought that if she could just get the cuff off her hand, she could escape. Idea after idea raced through her mind, but there was no foolproof plan. She was stuck there.

After a few minutes, the guy came back in and took her back to the chair. As he reached to clasp the cuff to the table leg, his hand slid across her, and he touched both of her breasts. She said nothing but a fear like she had never known arose within her. The guy said nothing as he hooked the cuff to the table. He sat across from her and she could tell he was staring at her. She turned sideways, trying to reveal as little of her front as possible. The man got up and went to a small counter and lit a propane burner. She watched as he opened a can of soup and put it into a pan. He pumped a handle by the sink and filled the pan with some water. He put the pan on the stove and leaned against the counter. She cringed when he spoke to her. "You are very attractive, but I suppose you already know that. I'll bet you and that boyfriend of yours have a good time." Emily said nothing and stared in a different direction. When the soup was warm, he poured a portion into a cup and put it on the table. He then took off the handcuffs. Emily was starving, so she drank the warm soup and put the cup back on the table. The guy took the empty cup and gave her a glass of water. She said thanks, but really meant she'd like to kick his ass. After she finished, he put the handcuffs back on, much to Emily's disgust.

At about six, the phone rang, and Aaron grabbed it. "Hello?"

"Hello, Aaron. Have you heard from her? Her mother and I are getting nervous."

"So am I, Don. I think I'll call the state police and see if there has been an accident on Highway 26."

"We've already done that, and there have not been any accidents. If we don't hear from her very soon, we're going to call the police and report her missing."

"I don't think they will give you the time of day unless someone's been missing for a whole day, but it's worth a try."

"Call if you hear anything."

"I will, Don. You do the same."

Emily could see the sun as it slowly dipped behind the mountain to the west of the cabin. The man lit a hurricane lamp and set it on the table. "We have some business to take care of." Emily swore to herself that she'd die before she'd let this guy rape her. He took a phone from his inside coat pocket. As he opened his coat, she saw the shoulder holster and wished she had the gun that was in it.

"Home number?" he said. Emily said nothing. "If you don't give me your home number right now, I will do unspeakable things to you." She reeled off the number so fast that he had to ask a second time.

He punched in the numbers and put the phone to his ear. "Yes, sir, I have your daughter. Now listen very carefully. If you call the authorities, she will die, and she will die a horrible death. Do you understand what I am saying? Good. Now I will let you speak with her for a moment. Remember, no calls to the cops." He handed the phone to Emily.

"Yes, Dad, I'm fine. Please tell Aaron I'm okay. Just a minute." She looked at her captor. "My dad wants to know if you want money?" He grabbed the phone from her hand and clicked it off.

"Some people are so stupid. What we want is what you little shits stole from the professor's house. Why don't you just tell me where it is and save us both a lot of time and trouble?"

"I don't know what the hell you're talking about," she said. "What professor?"

As he left through the front door, Emily reached into her bra with her free hand and grabbed the map. She could see in the dim light the parts that showed where both books were hidden. She took two big bites out of the map and examined what was left. One more small bite, and all of the crucial information was in her mouth. She folded what remained and stuffed it back into her bra. She grabbed the glass of water off the table and washed the paper down with the last gulp. She thought to herself, as the last bit slid down her throat, "Find that, you dirty son of a bitch."

Aaron grabbed the phone as soon as it rang. "Hello?"

"Aaron, listen very carefully," Emily's dad said. "Do not call the police. Emily is okay, but if the police get involved whoever has her says they will kill her. We found her car parked in the driveway. Do you have any idea what's going on? Do you have a cell phone handy?"

"I have one. Why?"

"Here's my cell number. Call me back on it." It wasn't more than a minute before Aaron called Emily's dad back.

"We have something those guys want. I'll gladly give it to them, but not until Emily is safe."

"What do you have that's so important that they would kidnap my kid to get it back?"

"That's not important, Don, but believe me. They want it."

"Aaron, I'm going to trust you, because I have no other choice. If anything happens to her, it will destroy her mother and me."

"Don, I'm scared shitless, but I'll die if that's what it takes to get her back."

Aaron asked Mary to go for a little ride. "Where are we going?"

"To re-bury the book."

"Why are we going to re-bury it?"

"They've kidnapped Emily, and I know what they want in return."

Aaron grabbed his phone and dialed Herb. Rachael was crying when she answered the phone. "Is Herb there?"

"No, he's been arrested for setting the ranger station on fire. They left here about a half an hour ago."

"Did they find the map?"

"No, they didn't even look for anything. They just read him his rights and left."

"Rachael, do you know if they've arrested James?"

"I don't know. Shall I call him?"

"Yes, and if he's still free, tell him to hide out somewhere and call me as soon as possible."

" Okay. Did you hear from Emily?"

"They have her, but not in jail. They've kidnapped her."

"Oh, Aaron, what are we going to do?"

"Just sit tight and don't give anyone any information, no matter what."

" Okay. Call me."

"Sit tight. I will"

Aaron's phone rang and he had it clicked on and to his ear faster than a gunslinger. "Hello, this is Aaron."

"What the hell's going on, Aaron?"

"Listen to me, James, and listen closely. Get in your car and start toward LaGrande. Go over Highway 58 and take the Burn's cutoff out of Bend. Call me when you get close, and I'll meet you. If you get stopped or arrested, don't give them any information. We're holding the cards, James. Do you understand me?"

The phone was silent, and Aaron had to ask the question again. Finally, James responded and promised to do as Aaron had asked.

Aaron didn't need the map to find the buried original. He had it dug up in a matter of minutes. They headed back to the ranch and went to the barn. Aaron grabbed the feed can and picked up a shovel. They got back in the truck and drove to the logging site where Aaron had killed the cougar. They went to the tree where he had saved the girls and dug a hole at the base. They placed the two books in the hole and covered them with dirt. While Mary held the light, Aaron spread pine needles over the fresh dirt and placed a flat rock over the spot.

When they were done, there was no sign that the earth had been disturbed. Mary questioned Aaron as they walked back to the truck. "Why on earth did we come all the way out here to hide them?"

"Because there is a third party who knows exactly where this tree is. I am going to make sure that if something happens to us, the books will be recovered."

Aaron drove to the mill, and they went into the office. Aaron dictated as Mary typed. "Officer Hunt, I trust that you remember me. I'm the guy who killed the cougar up by Trinity Butte. You know, the guy without a tag. I have buried something at the base of the cougar tree under a flat rock that is of the utmost interest to the government. Please retrieve it and deliver it to Senator Frank Jones. I'm sure he will be most appreciative. Thank you, Aaron Douglas." Mary called Dave and asked him to meet them in the high school parking lot.

Aaron gave the note to Dave and told him that if anything happened to Mary and him, he should deliver it to one J.D. Hunt of the Oregon Fish and Game department. Mary left with Dave, and Aaron drove home.

Emily could hardly sleep with her right arm and left leg anchored to the bed. She managed a few winks, but mostly lay there thinking of how to escape. Time ticked by slowly through the night.

Between being uncomfortable and having to listen to the loud snoring, she spent a miserable night. She thought of Aaron and her folks, and how worried they must be. At the first sign of light, the guy was up and making coffee on the small propane stove. On the second burner he cooked bacon and eggs. He gave Emily two pieces of bread and a cup of coffee as he sat back and enjoyed his breakfast. She said nothing, but ate both pieces of bread and drank the cup of coffee. After she had eaten, she asked to use the bathroom.

He said nothing as he unlocked the leg irons and the handcuffs. He again attached the cuffs to the pipe and left the room. Emily pumped the small foot pump to flush the toilet and then the one by the bathroom sink. She splashed water on her face and washed the best that she could. She took the dirty towel off the small protruding hook and dried her face.

Chapter XVI

"I don't want to know who you have. I just want that book back. Don't ever give me information I don't need. Now do your job." Marty Smith slammed the handset down, chipping a small piece of plastic off the phone. "Goddamned incompetent pieces of shit!" Marty walked through the den door of the governor's mansion and out through the sliding doors that led to the patio. His wife, Maggie, sat enjoying a cup of coffee in the morning sun.

"What's the matter, Marty? One of the sluts you're screwing call to tell you she's knocked up?"

"You know, Maggie, I can't wait until this term is over, so I can get rid of you and this fucking job."

"You love what you do, you sick piece of shit."

"I love you too, Maggie. I'll be back later."

Marty pulled out of the mansion but headed straight for Professor Finley's house. The shades were all pulled down, but the professor's car was in the driveway. Marty knocked on the door and heard noises coming from inside the house. Finally, the professor answered the door, and Marty stepped inside. "Gutsy move coming over here, Marty. Are you getting a bit bolder in your old age?"

"I'm not that old, Jonathon. That would be you."

"What brings you over here this fine morning?"

"I want to know what's in the book that was taken from you." Finley sat down on the sofa and didn't say anything for several minutes. He finally started telling Marty the story. He left out nothing. As he spoke, he could see the fear well up in Marty's eyes.

He took pleasure in explaining every detail of how he had persuaded and encouraged certain special students to become soldiers of destruction. He mentioned Gregg Summers, the two-term governor who started the land use planning that virtually took property from private citizens and left the state in total control. "You see, Marty, you have been designed to perform your duties in a specific manner. You, as Gregg Summers had done, did a magnificent job of crippling the economy of this state. And the beauty of it is, it has spread to other states and even other countries. It's all in the book, with names and places penciled in for your reading enjoyment."

Marty got up and walked to the door. "You're telling me the truth, aren't you? You've done all this "let's save the world" shit just to fuck up everything. I hope we find the fucking book, but not to save your sorry ass. You better book a ticket to the Ukraine."

Marty left and went to the state police facility where Herb was being held. The guard on duty jumped to his feet when he saw the governor approach. "Good morning, Governor Smith. What brings you here?"

"I need to have a private meeting with one of your detainees. Could you get Herb Snyder and give us some time in a private room?"

"Yes, sir. Why don't you go down to the conference room? It's just down the hall on your right. I'll bring Snyder down to you."

"Thank you."

Herb couldn't believe who was waiting for him in the conference room. "Mr. Snyder, I'm Governor Smith. I'd like to ask you a couple of questions."

"By all means, Governor Smith. I take it they are important questions?"

"I know you have knowledge as to the whereabouts of a certain book. I think we both know what book I'm talking about. I know you don't want to go to prison for burning down the ranger station, so I thought we might make a trade."

"I don't think I'll go to prison for burning down the ranger station, because I didn't do it, and I have a witness. And as for a book, I have no idea what you're talking about."

"Then let me ask you if you'd like to end up like your friend Ernie?" Upon the threat of being murdered, Herb stood and yelled at the top of his lungs.

"Help me! the governor has threatened to have me killed. Help me! Help me! Please, help me!" Marty stood and approached Herb.

"Shut up, you fucking little shit! I'm not going to have anyone killed." Herb kept screaming, until two guards entered the conference room. One of the guards grabbed Herb and told him to shut up. The other asked the governor what was going on. The governor told the guards that he thought Herb was crazy.

Herb spoke to the guard who was holding him by the arm. "I'm not crazy. The governor here told me that he was going to have me killed, just like someone named Ernie. You might want to check it out; that name, again, was Ernie."

Marty Smith stormed out of the conference room and through the front door of the holding facility. He was so nervous that he forgot where he had parked his Lexus. He finally found it and began driving, but not toward the mansion.

He drove west on Highway 126 and didn't stop until he had reached Honeyman State Park, just south of Florence. He parked the car and walked out on to the sand dunes that separated Highway 101 from the ocean. The only thing on Marty's mind was how to save his own ass. Threatening the little shits that stole the book might not be such a good idea, especially now. His only hope was to trade the girl for the book, and to do it now.

Chapter XVII

Aaron got the call just after noon on Sunday. He had talked A.J. into letting him work as day watchman at the mill, instead of helping Jed.

"Hello?"

"Hi, Aaron. It's James. I'm just outside of LaGrande at some place called Ukiah."

"You're still an hour away. I'm working at the mill until three. Cool it for a while and meet me at the Chevron station at the Eastern College exit at about ten after three. If you leave Ukiah at about two, you won't have long to wait."

"Okay. Have you heard from Emily?"

"I'll talk to you when you get here."

Aaron called Mary, who was manning the phone at the ranch. "Anyone call yet?"

"I'll call you as soon as anyone calls. Don't be so jumpy. You said yourself that we are in the driver's seat."

"I know. I'm just worried about Emily. If anything happens to her, it will be my fault. We should have left it alone."

"Nothing is going to happen to her. She's tough."

"As soon as they call."

"Okay. Don't worry."

Aaron sat down out by the log yard and counted the trucks in the truck shop. He thought to himself that there was nothing more important than getting Emily back in one piece. He then wished he'd turned the book over as soon as he'd found it. But he had no idea who he could or couldn't trust. The book was an incredible find, and he would have loved exposing the professor for what he really was, for what they all really were.

At 3 P.M. sharp, Aaron's replacement pulled into the parking area. Aaron jumped into his truck and waved as he headed out of the mill toward town. James was parked alongside the service station, and Aaron pulled up beside him and motioned for James to follow. He led James to the ranch and had him park his car out behind the barn. Aaron filled James in on the Emily story as they walked to the house.

A.J. and Nancy were on a two-day log-buying expedition over toward Baker. They wouldn't be back until some time on Tuesday and Aaron hoped this would all be over by then. Aaron introduced James to Mary. "So this is the hotshot typist who helped translate the book. Who was the other kid who did the actual translation?"

"His name is…" Mary began.

Aaron stopped Mary in mid-sentence.

"Doesn't matter; let's just say he did a good job and leave it at that."

Just before five, the phone rang. Mary answered. "It's for you." She handed the phone to Aaron.

"Aaron, this is Don. I've just had a phone call from the people who have Emily. They told me that her release depends entirely on you turning over some book that you stole from some guy. Do you have the book?"

"I can't say; I might. If I do, no one will see it until Emily is safe. Do you hear that you people that are tapping my phone? Not until she's safe. Goodbye, Don."

Aaron hung up the phone and then waited for the call. It came just after seven. The voice on the other end was as cold as ice. "We will tell you what to do with the book. If you don't do exactly as we say, we will kill the girl, and she will die a most horrible death. Bring the book to the county courthouse, and place it in the mail drop. If it isn't there by ten this evening, the girl dies. Do you understand?"

"Listen to me, you ignorant fuckhead. If Emily Alcott isn't released into the custody of her parents by eight tomorrow morning, that book will be in the hands of the Republican senator from the great state of Oregon, with copies going to all of the major news stations. We have this conversation recorded and will send a copy along with the book. Your move."

There was a long pause and finally an answer. "You be there, at the girl's parents' place with the book, at 8 A.M. sharp. The girl will be there, unharmed. If you call the police, or try anything else, the girl won't see another day."

"I'll be there." Aaron hung up the phone and jumped into the air. James, caught up in the moment, jumped up and gave Aaron the old high five.

When James lifted his arms, in what was meant to be an act of jubilation, a small transmitter fell out from under his shirt and dangled from a small black wire. Aaron saw the small black device and immediately grabbed it. James was shaking so hard that he nearly fell on the floor. Aaron put his index finger over his mouth, telling James to be quiet. He went to the desk in the den and took a pen out of the desk drawer. He then started telling Mary and James how he had outfoxed the pros on this one. While he was bragging, he wrote a note to James. "Who wired you? Are you one of them?"

James wrote back. "No, they threatened to kill my mom and me if I didn't cooperate. I'm sorry, Aaron. I didn't know what to do."

Aaron wrote back. "Don't worry about it. Just follow my lead." James nodded his head, acknowledging that he understood.

"First thing we need to do is get the books dug up. They are buried out by the old pioneer cemetery; we thought that was a fitting place. Mary and I will go dig them up; you stay here and watch TV. Whatever you do, don't answer the phone." Aaron turned on the TV and lifted James' shirt up over his head. He carefully took the transmitter off of James and laid it in a chair by the TV. The three of them left and headed for the cougar tree.

"So where were you when they wired you?"

"Just this side of Bend. That's why it took me so long to get here. Aaron, do you think they will kill my mom?"

"They won't kill your mom." They drove to the landing and all three walked to the tree. Aaron dug up the books and carried them back to the truck. "It's a long-ass drive to Portland. I hope those bastards have her there like they're supposed to."

It was just after nine, and the guy had just given Emily another cup of soup. She could smell the steak that he was cooking for himself and thought about what a jerk he was.

She hadn't been to the bathroom for hours and asked if she could go. She was flushing the toilet as the guy's satellite phone rang. She opened the door just as he answered. "Eight-forty go. Tonight? You've got to be shitting me. That's over a 200-mile drive. Okay. I'm gonna finish dinner and then I'll head out. Don't worry. I'll have her all taped up for the occasion." He took her from the bathroom and sat her in the chair. This time, he put only the handcuffs on her and left the leg irons wrapped around the chair.

"Good news. You're going home. Good thing, too, because I was getting really horny looking at you Thank God for that peep hole in the bathroom door." Emily sat still and said nothing. She finished her soup and watched in amazement as the guy devoured his steak in record time.

His phone lay just beyond her reach, and she thought of how she would like to talk to Aaron and her folks. She thought of her nephew, Mikey, and wished she could hold him. Just as the guy was finishing his last huge bite of steak, he started choking. The big piece of steak had lodged firmly in his throat, and he could neither inhale nor expel air. Emily just sat there in amazement as he fought to breathe. She could see the fear in his eyes. "Quick," she said. "Give me the cuff key, and I'll give you the Heimlich maneuver. I'm trained in First Aid. Give me the key."

He reached in his pocket and pulled the key out. He handed it to her, and she quickly unlocked the cuffs and moved around behind him. She put her left arm around him and placed it just under his sternum. She took her right hand and put it around him pulling the revolver from his shoulder holster. She pushed him as hard as she could, and he tripped over her chair and landed face first on the floor. She pointed the revolver at him and pulled the hammer back. She couldn't fire the weapon and laid it on the table. She turned the guy over and looked into his dimming eyes. His face was turning blue and somehow she felt just enough compassion to stand on the table and jump onto the guy's stomach. A loud rush of air dislodged the piece of meat and set it flying into the air. He was nearly unconscious, but she could see color returning to his face.

She grabbed the phone and the revolver from the table and headed for the door. She knew that she should kill him and get his car keys, but something inside her wouldn't let her do it. She ran from the cabin and, in the dwindling light, managed to shoot out both of his front tires. She could make out the road and thought that it headed east. She had no idea where she was, but didn't hesitate on which way to go. She headed away from the road across a small clearing and into the trees. She was sure the guy in the cabin wouldn't cause her any more problems, but how soon would they have reinforcements?

When she had walked for about an hour, she stopped and took the phone out of her pocket. She pushed one of the buttons and the light on the face came on. She dialed Aaron's cell phone but couldn't get it to connect. On the face of the phone she could see the words

"Enter Code." She threw the phone and continued walking through the woods.

There was a partial moon, but clouds intermittently blocked what little light there was. She walked in what she thought was a straight line, moving faster when there was moon light and more slowly when it was totally dark. She kept the revolver in her right hand and her thumb on the hammer.

After she had walked for a good four hours, the terrain started to change. She was walking downhill, and it was getting steeper and steeper the further she went. She fell twice, the second time hitting her hip on some rocks. She was in an area where there was a lot of shale rock, and it was getting harder to navigate. She knew that they probably wouldn't even know she was gone until daylight. They would surely wait for the guy to deliver her, and when he didn't show on time, they would send someone to look for them.

She had time on her side, so she would wait until morning and not risk an injury that could slow her down, or even stop her altogether.

It was just past midnight by the time they had retrieved the books. They took Mary's VW and headed for Portland. Aaron couldn't resist driving past the old cemetery. There were two cars parked about a half mile from the cemetery, and as they drove by, the headlights on one of the cars came on. "Aaron, this was a bad idea." Mary was in the backseat with her eyes glued to the headlights about 200 yards behind them. Just past the cemetery, the car pulled off the road and turned its lights off.

Mary breathed a sigh of relief. "They stopped. Do you think it was them?"

James answered, "I think that was the same two guys that wired me. The car sure looked the same." They drove the back roads to Perry and took I-84 West. It was after one in the morning by the time they were on their way. Aaron had made the drive in five hours, but this time there would be no speeding. They didn't need a reason to be pulled over and possibly searched.

The trip proved uneventful, and they hit the city limits just before seven. They stopped at a truck stop and got some coffee and rolls. They filled the gas tank and drove west toward Tigard. Aaron called Emily's parents and Don answered.

"Hi, Don. They are supposed to have Emily there by eight. If they do, call me on my cell. Do you still have the number? Good. Tell them we'll be there in 10 minutes." Aaron drove to the Denny's on I-5 and pulled to the very end of the parking lot. He turned off the motor and waited.

There had been enough light to start walking at 5 A.M., and Emily had made the most of it. By 7 A.M., she had traversed what had been a very steep mountain and was now in a valley. There was a good-sized stream running through it. She walked downstream and thought about getting in the water to throw off any would-be bloodhounds that might be following her. She was very thirsty but had heard enough stories about Beaver Fever to resist the temptation.

She had walked for about an hour when she saw the most welcome sight of her otherwise miserable adventure. There, on the other side of the stream, was a dirt road and a utility pole with wires attached. She crossed the stream and the cool water felt good as it seeped into her tennis shoes. Once on the road, she felt like running, but decided a fast walk would serve her better.

At 8 A.M. sharp, Aaron's phone rang.

"Hello? Is she there?"

"Not yet, Aaron. Are you sure they said eight sharp?"

"That's what the guy said. I hope this wasn't some kind of a set-up."

"Me, too. I'll call you as soon as I hear something."

"I won't give them what they want until Emily is safe. Tell them—no Emily, no book."

"What book, Aaron?"

"Just tell them that." Aaron clicked the phone off and laid it on the seat. He sat and thought to himself, "No matter what, they get nothing until Emily is safe."

The helicopter set down not 50 yards from the cabin. Emily's captor, Officer Oren Oswald, sat on the small porch and held his head in his hands.

"What in the hell happened, Oren?"

He didn't answer, and Officer Millbright became agitated. "Look at me, you stupid asshole. Where is Emily Alcott?"

"If you call me a stupid asshole again, I'll shove my fucking fist through your skinny fucking head, do you understand me?"

"Settle down, Oren. What happened?"

"I was choking, and that kid saved my life. She didn't have to, but she did. If we find her, I'll do what I can to help her, and the fucking governor can kiss my ass."

"That's just great, Oren. We work for old Marty boy and like it or not, we do what he tells us to do."

"Not any more Millbright. From now on, I'll do my job, but I won't break the law. Refresh my memory, Millbright. Is kidnapping against the law?"

Emily saw the tiny cabin and approached it with caution. There were two wires running into the cabin, and Emily figured there must be a phone. She knocked on the door, and no one came. She tried all the doors and windows, but they were secured. She found a rock and was about to break a window when she noticed there were no furnishings in the cabin. She went to each window and looked for a phone. She saw the phone jack, but there was no phone. There was a water faucet by the front porch and as she turned it on pure fresh water flowed from it. She drank until she was satisfied but didn't overdo it. She kept the rock in her hand, shoved the pistol down the back of her pants, and pulled out her tee shirt and let it hang, concealing the gun. She continued down the dirt road; there would surely be another cabin.

At a quarter 'til nine, Aaron's phone rang again. "Hello? What's up?" It was Emily's dad.

"Aaron, there is a gentleman here who would like to speak to you."

"Who is he, Don?"

"I'm not sure. Hang on."

"Aaron, my name isn't important. There has been a change of plans. Before we can turn Miss Alcott over, we will have to have you return our book. Bring it to the Alcott residence now, and I assure you the girl will be returned to her family unharmed."

"Where is Emily? We had a deal."

"We have decided that what we have to offer is of greater importance than what you have. Wouldn't you agree?"

"I don't think you give a shit about Emily, but I know what's in the book, and it makes all you fucking liberals look like idiots. If you don't have Emily back to us in exactly 30 minutes, everyone in America will know how Marty Smith and Senator Driggers got elected. If you don't want that information leaked to the public, give her back." There was a long pause.

"Don't do anything you'll be sorry for. I'll call my superiors and get back to you shortly."

Emily was hungry, but otherwise in great spirits. She was worried sick about what her parents and Aaron were thinking, but for now she had to think of herself. She walked until just after ten before she saw any more signs of life. Another cabin, and this time a car was parked in the tiny dirt driveway. She cautiously approached the cabin and knocked on the door. She was relieved when an elderly lady answered the door. Emily asked if she might use the lady's phone, but became disheartened when the lady explained that she had never had a phone installed. "I love the peace and quiet up here. I'm not about to spoil it with a phone ringing off the hook."

"Where is 'up here'?"

"Why, my dear, are you lost?"

"I think I'm just disoriented. Maybe you could point me in the right direction."

"Well, if you go another two miles down this road, you'll come to a paved road. If you turn left and go another seven miles, you'll come to Highway 31, just south of Silver Lake. Now do you know where you are?"

"I'd really like to say yes, but the truth of it is, I have no idea where Silver Lake is."

"Have you ever been to LaPine?"

"Yes, not too long ago. Are we close to LaPine?"

"About 50 miles from here. Would you like a ride?"

"I'd be more grateful than you could ever imagine."

"I'll give you a ride to LaPine, but only after you've eaten something. Come in and have a seat."

No one spoke after the phone call. Aaron sat, motionless, and stared through the windshield at a large laurel hedge that separated the

Denny's from business next door. They waited until almost 10:30, and then finally the phone rang. "This is Aaron."

"Mr. Douglas, this is Governor Smith. We have a terrible situation here. You have something I want." Aaron knew that they had gotten to Dave, and that he must have cracked. Thank God that they had dug the book up and had it with them.

"We do have a terrible situation here, but all you have to do is give Emily back, and you can have your book. I have it here with me. Where is Emily?"

Marty Smith was desperate. He had no hostage, so he decided to take a chance. "Aaron, listen to me. I know that the most important thing in the world to you is your friend Emily. Now here is the deal. If I don't have the book in my hand by noon, she will meet with a most unfortunate accident. She will be killed while resisting arrest for the burning of the ranger station in Oak Ridge. You and your friends will go to jail for a long time for the same crime. If you want to make the book public, go ahead. My term of governor has only 14 months left. You'll receive a phone call at 11 A.M. telling you where to deliver the book"

"Where is Emily?"

"Noon, Mr. Douglas."

Mary put her hand on Aaron's shoulder. "What kind of a mess have we gotten ourselves into?"

"No big deal, Mary. That was just the governor threatening to kill Emily and put the rest of us in jail."

James put his hands on his head and started swaying back and forth. "This is so stupid! We should never have gone to the professor's house." Aaron grabbed his right hand and squeezed it as hard as he could.

"Knock it off, Aaron. That hurts," said James.

"James, stop whining. We're in a situation that whining won't fix." They sat there in silence for a few minutes.

Aaron's voice was shaky when he spoke. "Let's give it back. It doesn't sound like we have another choice."

James agreed with Aaron, but Mary didn't like the idea. "If we give them the books, they have everything, and we have nothing. I think we should keep the books until they give us Emily."

Aaron put his hand on either side of Mary's head and turned it toward him. "He said they would kill her if the book wasn't back by noon. We're giving the books to the assholes, like it or not."

Emily was surprised at how fast the lady drove. She had gulped down the scrambled eggs and toast, and they hit Highway 31 just before eleven. "How long will it take us to get to LaPine?"

"About 45 minutes, barring any unforeseen situations." Emily thought to herself that there couldn't be a situation as unforeseen as what she had just been through.

"I'd like to get your name and address so I could send you something."

"Don't worry about that, dear. I wasn't busy, and I needed to get groceries anyway. Do you have money for a phone call?"

"I have nothing. No purse, no credit card, nothing." The lady drove with her left hand and fished around in her purse with her right.

She pulled a wallet out of her purse and handed it to Emily. "There should be a twenty in there; that may get you by for a while. Do you have anyone you can call?"

"My parents and my boyfriend."

"You have parents and a boyfriend? How on earth did you wind up lost in the mountains?"

"I would tell you, but it is probably best if you don't know." The lady didn't ask any more questions.

At eleven sharp, Aaron's phone rang. "Bring the book to the radio tower on Council Crest. Leave it on the ground by the gate to the enclosure. The girl will get in touch with you after we have the book in our possession. Do you understand me?"

"Yes, I understand."

"Remember, Mr. Douglas, if you do not do exactly as you are told, she will die a horrible death." Aaron clicked the end button on his phone and set it on the seat. He asked James if he knew how to get to Council Crest, and he didn't.

He picked the phone back up and dialed.

"Hello?"

"Hey, Robby, this is Aaron."

"Hey, Aaron, how are you? Still got that good-looking girlfriend?"

"Yeah, Robby, I do. Do you know how to get to Council Crest from I-5?"

"I go to school up there, Aaron; of course I know." Robby gave Aaron directions, and Aaron jotted them down on a small piece of paper.

"Thanks, Robby. Are you in school now?"

"Not for a couple more weeks. Why?"

"I might need some help."

"Aaron, what the hell's going on?"

"I'll tell you later."

"Call if you need me."

"Thanks, Robby."

It was a quarter to twelve when the lady dropped Emily off at the truck stop on the north end of town. "I can't thank you enough; you're a life saver."

"That's all right, Honey. Good luck." Emily went into the truck stop and went to a pay phone.

She dialed her parents' number and waited for the operator. "What billing, please?"

"Collect from Emily Alcott."

"I have a collect call from an Emily Alcott. Will you accept the charges?"

"Yes! Emily, is this really you?"

"It's me, Dad. I'm in."

"Don't tell me they let you go. I thought they were lying."

"Lying about what?"

"About letting you go if Aaron turned their property over."

"They didn't let me go. I got away. It's a long story, but a good one. I'll call you back. I better get a hold of Aaron."

Aaron parked the car and looked all around but couldn't see anyone. He was about 30 feet from the enclosure. He grabbed the bag with the books in it and walked toward the woven wire fence. He placed the bag on the ground by the gate and went back to the car. As he started to turn around, the phone rang. "They're where they're supposed to be. Now where is Emily?"

"Right here, Sweetheart. Where are you?"

"They let you go already?"

"Not hardly; I escaped last night."

"Are you safe?"

"Yes. Why?"

"Call me back in 10 minutes." Aaron put the car in reverse and backed as close as he could to the place where the books were lying.

"Mary, go grab our books."

"What's going on, Aaron? Is Emily all right?"

"Please, Mary. Just do it." Mary ran to the bag and picked it up. Aaron drove slowly away and headed down the hill toward Portland. About a quarter mile down the road a green sedan passed them, driving slowly up the hill. The driver had a huge smile on his face when he saw them. James was looking directly at the driver and gasped when he saw who it was.

"Holy shit, that's one of the guys that made us fill up the gas cans."

"Are you sure?" asked Aaron.

"Very sure. When someone makes you piss your pants, you don't forget them." As soon as the car was out of sight, Aaron hit the accelerator.

The VW bolted down the hill. Aaron had no idea where he was, or where he was going. In a matter of minutes, he was at the bottom of the hill. He drove into a housing area by the river and parked on a side street. Aaron grabbed his phone and called Robby. "Robby, it's me again. Are you in for some excitement?"

"What kind of excitement?"

" Real-life cops and robbers."

"What shall I do?"

"Meet me in front of Martha's house in 20 minutes."

Emily called back, and Aaron explained the situation, leaving out the part about them threatening to kill her. He listened intently as she told him about her imprisonment and her escape. "I'm impressed! I'd never have found my way out of the woods. Why don't you go to the place where we ate and wait for us?"

"I can do that. How long are you going to be?"

"Give us about four hours. I don't want to get stopped for speeding."

"I'll be there."

Aaron ran into his grandmother's house and found Martha in the kitchen, sweeping the floor. "Hi, grandma, Grandma! Can I borrow your van for a couple of days? I don't have time to explain, but it's really important."

Martha gave Aaron the keys to her van and followed him outside. Martha saw Mary as she was getting out of the VW and gave her a big hug. "What in the heck is going on? Are you guys in trouble?"

Mary answered, "Grandma, all I can tell you is that we are the good guys. You'll just have to believe me on this one." Aaron gave Martha the VW key and waited for Robby. The fancy Porsche pulled around the corner and into the driveway. Aaron handed Robby the bag and asked him to hide it and not tell anyone where it was hidden. Robby took the bag and was standing there, asking questions as Aaron, Mary and James were getting into Martha's van.

"Robby, this is important. I can't tell you about it now, but I'll call and fill in the blanks. Thanks a million; I owe you one."

"Whatever you're doing, Aaron, I wish you luck." Aaron waited for Robby to move his car and then backed slowly out of the driveway and drove south out of Portland.

Roderick Larios

Chapter XVIII

Marty Smith was livid. He ranted and raved, shouting obscenities at Oswald and Millbright. "I can't believe you let her get away. And now Johnson tells me the kid stiffed us on the book. I'm fucked, totally fucked."

Millbright stood and walked toward Marty. He was trembling, and his voice was very shaky when he spoke. "I don't know what's in the book that you so desperately seek, but I'm pretty sure it doesn't have anything to do with Oswald and me. I've always tried to do a good job for you, and I did it knowing some of the things I was doing weren't right. I'm done breaking the law. If you want me to work for you, I will, but not if it means breaking the law."

Marty turned away from Millbright and spoke in a more controlled tone. "Don't be so righteous, Officer Millbright. Some of the worst things we ever did were your ideas. None of us has been saintly. I want you to get the book back from those little bastards before something really bad happens. Do this one last job as you see fit, but please do it. How hard can it be to take something away from a bunch of kids?"

Millbright and Oswald reluctantly agreed to give it another try. They left Marty alone in his office. Marty called Eric Johnson and ordered him to get in touch with Millbright and Oswald. He trusted Johnson more than any of the other six aides he had enlisted to do his dirty work.

Oswald and Johnson had burned the ranger station down after implicating Herb and James with the gas cans. It wasn't the first horrific act of vandalism they had performed. Over the years, there had been several other acts committed in the never-ending struggle to save the environment.

Marty rarely got involved with the details. He let Tom Erickson plan and carry out most of the missions, staying far enough away from the action to keep his hands clean. He was tired and near the end of his term. All that Marty Smith wanted was out. Just recover the book before it went public, and everything would be fine.

Marty stared at the ringing phone on his desk. Finally, he picked up the receiver. "Sir, there is a man on the phone by the name of Aaron. I told him you weren't available, but he insisted that you would want to talk to him."

"Put him through." Smith cleared his throat. "Hello, Aaron. You didn't do what you were supposed to do. You are in big trouble. You'll be arrested and sent to jail for burning down the ranger station if you don't give back our property."

Marty Smith listened in horror as Aaron spoke. "We know who the two guys were that burned the ranger station down. They work for you. Now I've got a deal for you. Let Herb Snyder out of jail and drop all charges, or the biggest story to ever hit the coast will be on the evening news. You should see the notes in the original. There are several Russian words and then the name Marty Smith and then some more Russian words, and then Marty Smith again. You seem to be the star of the book. What's really interesting is the translation. The old boy just sat up above you and pulled all the strings, and you did everything that was asked of you. You have my cell number. Have Herb call me within the hour."

"Listen to me, you little shit. I don't have the power to get someone out of jail by just asking."

"Listen to me, you stupid shit. One hour."

Aaron, Mary and James had just driven through Sisters as the phone rang. "Hello?"

"What the hell did you do, Aaron? I've seen pissed-off people before, but Marty Smith is one step beyond that," Herb said.

"Good. Are you on a cell phone?"

"Yes, and its Rachael's. We're safe."

"Herb, we need another meeting. Can you come to LaGrande?"

"Sure, we'll be there as soon as we can. Where should we meet you?"

"Register at the Motel Six under the name of Bob Jones. We'll meet you there at eleven this evening."

"Hey, Aaron."

"Yeah?"

"Thanks for springing me."

"Hey, Herb?"

"Yeah?"

"We have to figure out what to do with this thing."

"We will. See you tonight."

Emily sat in a small tavern and drank a Diet Coke. There were two guys playing pool and the bartender. The tavern was two blocks off the main highway, and Emily felt safe there. The bartender was busy washing beer glasses, and the two guys playing pool were engrossed in their game. She sat there thinking about what a serious situation they were in. She was glad that Aaron still had the book, but didn't have any idea what to do with it. What reaction would people have? Would they think it was real, or would they say it was a hoax?

She wished that they had never found the book. All she wanted was her life back.

Herb and Rachael drove east over Highway 58. The warm afternoon sun forced them to open the windows of Herb's Buick. "I really need to get the air fixed on this thing."

"We could have taken my car."

"Yes, Rachael, we could have, but we didn't."

"What are you so bitchy about?"

"I'm sorry. I'm trying to come up with some super idea about what to do with that book. I want the world to know how all of us fucking liberals materialized, but I just can't think of a good way to present it."

"Why don't we just give it to the police and forget about it?"

"We need to think of ourselves a little bit here. Both of us could go to jail for a long time. If we bring Marty and his buddies down, we'll most likely be right along beside them."

"Are you suggesting we make a deal?"

"I'm suggesting that making some kind of a deal is a possibility."

Aaron, Mary and James discussed the different options. Aaron and Mary were leaning toward giving the book to the authorities, and James was casually mentioning the possibility of trading the book for a clean record.

Aaron stopped for gas on the north side of Bend. They all used the restroom and all bought a snack and a drink. As they were pulling out of the station, Aaron spoke. "Tell me, James. After all we've been through, would you give that book back to those people?"

"I don't know, Aaron. Keep in mind that there are pictures of Emily pouring gas on the ranger station. I just wonder who all will be hurt when the shit hits the fan."

"Well, you have a right to your ideas, just like the rest of us. After all, James, you're the one who found the book."

"I don't want to give it back. It would be great seeing Jonathon Finley and Marty Smith squirm a little."

Marty Smith drove to the governor's mansion the long way. His mind was going a hundred miles an hour. As he walked through the huge oak doors, Maggie and her therapist emerged from the master suite. "Well, this is cute," Marty said. "Some special therapy today?"

"No, Marty. As much as you'd like me to be screwing Kevin here, it was just a lesson in dealing with my fears."

"Afraid of the bedroom?"

"No, Marty, afraid of what's in the bedroom."

"Why are you always so dramatic? You know, Maggie, you just aren't very smart." Maggie went back into the master suite, and her therapist left through the opened front door.

Marty Smith sat in the middle of the huge front room and thought about what kind of a legacy he would leave behind. Would it be that of a caring and thoughtful person who had the interests of the common man at heart, or that of an accomplice to the most damaging group of individuals to ever take up arms against America?

Roderick Larios

Chapter XIX

Emily spent the last hour in LaPine walking through the rural neighborhoods. She made it back to the restaurant 20 minutes before the four hours were up. She didn't recognize the van and sat on a bench out front. Aaron saw her and nearly sprinted to her.

"God, you look good! I love you so much."

Emily threw her arms around him. "I love you, too. And you can't imagine how glad I am to see you. I've never been that scared."

"Did they hurt you?"

"No, just my pride. I got sucker punched. I never saw it coming." Mary and James both got out of the van and hugged Emily. Aaron suggested that they get rolling. They climbed in the van just as a state police officer turned the corner. Aaron held his breath as the guy drove by and waved. Aaron waved back and breathed a sigh of relief.

The drive to LaGrande took four hours, and Emily slept the entire way. They pulled into Sam's just before nine. Aaron's phone rang, and A.J. told him that he and Nancy wouldn't make it back from Baker until some time Wednesday morning. Aaron assured A.J. that everything was fine and said goodbye. They went into Sam's and ordered a late dinner. Aaron started laughing when Emily ordered roast beef and mashed potatoes. "You must be really hungry."

"I am. I sat and watched that asshole stuff himself while I ate runny soup. I guess it was a good thing, though, because half the time all I could think about was food. The lady that gave me the ride made me some scrambled eggs. It might have been the best thing I've ever eaten." They finished dinner and drove to the Motel Six. James was the first to recognize Herb's Buick.

"They're here. I'll call and see what room they're in."

Aaron dug a piece of paper out of his pocket and handed it to James.

"Call Rachael's number. I think it's safer than calling Herb's phone." James dialed, and Rachael answered. James got the room number, and Aaron pulled the van in next to the Buick.

Herb answered the door and let them into the room. His first order of business was to hug Emily. "I heard you had a bit of an adventure."

"It was a bit more than an adventure. It was a true nightmare." Herb shook Aaron's hand and thanked him again for getting him out of jail.

Aaron sat on the bed and started the meeting. "I think the first thing we should do is get an idea from everybody. Don't hold back to spare feelings. Tell the rest of us exactly what you would like to have done with the book. Since I'm already talking, I'll go ahead and give you my idea first. I think we should turn it over to the government and let them decide what should be made public. I don't really trust a lot of government officials, but I'm sure we could find someone who would do the right thing with this information. Whatever we decide to do, we need to do it quickly. There are people who will stop at nothing to get the book back. Mary, you're in this just as much as any of us. What's your idea?"

Mary didn't say a word for over a minute. She finally sat on the bed by Aaron and started talking. "I think we should give it to the media. Any one of the big three would plaster this all over the six o'clock news. We could keep the translated copy and give them only the original. I don't want to sound like a mercenary, but I'll bet this thing is worth a fortune. The only reservation I have about the mainstream media is the fact that they are one of the entities that were manipulated. I vote we try to sell it to the highest bidder and donate all the money to displaced timber workers."

Herb and James looked at each other but didn't say anything. Rachael bit her tongue to keep from saying something. Though her ideas were changed, she still didn't have a soft spot in her heart for timber people.

James spoke next. "I would like to see this information made available to the public. Though Russia no longer poses this kind of a threat, what they did devastated this country. I was blind when I chose to partake in all of those destructive acts. We were all blinded by the words of Professor Finley and so many others. Having said that, I would vote to turn the book back over to the governor and Finley in exchange for complete amnesty. I can't speak for the rest of the FF gang here, but I know that for what I did, I could spend years in jail. I never hurt anyone, but I easily could have. God help me for saying this, but I'd like to make a deal."

Herb stood up from the small table and walked to the door. He opened the door and stuck his head out, taking a deep breath. He shut the door and turned to the rest of the group. "Do you know what that is? It is the smell of freedom. I've spent the last few days locked up like an animal in a small cage. I really don't want to do that anymore. Aaron has proven that the book we have has great power. It got me out of jail with just a phone call. Like the rest of you, I would like nothing better than to see this information released, but not if it means spending the rest of my life behind bars. I vote we give it back in exchange for our freedom." Herb sat back down at the table and nodded at Rachael, who then stood to talk.

"I've given this a lot of thought. What I'd really like to do is finish school and get a job and forget this ever happened. On the other hand, I think what they did to us was horrible. We were used like little chess pieces to carry out very evil tasks. As much as I'd like to be free from the threat of incarceration, I am going to abstain from the voting.

I guess it's two to two, with Emily casting the final vote. Go ahead and tell us what you think. Don't worry. I won't change my mind and vote after you."

Emily walked over and kissed Aaron on the cheek. "I love you, but I'm going to have to vote against you this time. I want my life back. No, I want our lives back. If there is a way we can swap the book for complete amnesty, I say we do it. I think what they did was as evil as evil gets, but remember; a bunch of us went for it. Nobody held a gun to anyone's head. Even if this were exposed, it wouldn't change the way people feel. Liberalism is ingrained in people. It would be newsworthy, but it would pass just like everything else. If I thought there was even the slightest chance they were still sending people here to mess with our minds, I'd say expose them, but we know they're not. Let's get this resolved. It's a chance for those of us who could be in big trouble to have another chance."

Aaron put his arm around Mary. "Well, Sis, looks like we're outvoted. Are you okay with this?"

"Do I have a choice? We went through a lot of trouble with these books just to give them back. Am I okay with this?" Mary stood and walked over to Emily and put her arm around her. "If it means getting you for a sister-in-law, and it gets you out of hot water, I'm okay with it."

"So now what?" Herb asked. "Where are the books?"

Aaron stood up as he answered Herb. "They're in Portland, and they're safe. How do we go about getting amnesty for you guys? I guess I should rephrase that to include Mary and me, and probably Dave. Mary, did you see a card from a lawyer in Portland? It was in one of my shirt pockets."

"It's in the catchall drawer by the washing machine. Why?"

"I think we'll need someone to draft a contract for the governor to sign. That guy thinks he owes me a favor; maybe he'll be able to help us."

They devised a plan. Aaron would be in charge of getting a resolution drafted, and Herb would contact the governor and negotiate a trade. It wasn't what everyone wanted, but at least everyone was in agreement. Mary called Dave and found out that Marty's boys had found him by going to see Dr. Howley. He had told them about Mary's visit and gave them Dave's name. They had threatened Dave, telling him they were going to deport his mother if he didn't tell them everything he knew. Dave didn't tell them about the translation or show them the note. They had searched his car and found the note. Mary briefly told Dave of their plans and as she hung up the phone, Aaron overheard her say, "I love you, too."

Roderick Larios

Chapter XX

Herb and Rachael were home by two on Tuesday afternoon. Herb laid his notes out and called the governor's office. The receptionist put Herb right through to Marty.

"Hello, Herb. Calling with some more threats, I presume?"

"You presume wrong, Governor Smith. I'm calling to propose a deal that will be beneficial to all concerned. If you sign a document giving some of my friends and me complete amnesty, we will return your book and the translated copy, and promise not to divulge anything we have learned."

"Well, Herb, I'm glad you've come to your senses. I will have such a contract drawn up for my signature."

"That's all right, Governor. We are having the agreement drawn up as we speak. I will call and let you know when it is ready."

"Very well. I will wait for your call."

"Oh, by the way, Governor. It wouldn't be a real phone call without a little threat. If we see any sign of any of your boys bothering us, the book goes to the media immediately."

Herb hung the phone up and looked at Rachael. "That was the most fun phone call I have ever made. I think I'll go into politics."

Aaron, Emily, and Mary said goodbye to James and headed for Portland. Aaron had called the office of Howard Burkiwitz and made an appointment to see him. Ann's dad had remembered his offer and was pleased that Aaron had called him. They had a meeting scheduled for 4:30, giving Aaron and Mary enough time to take Emily home and drop the van off at Martha's.

Don and Trish were both crying as they hugged Emily. Don shook Aaron's hand and thanked him for bringing Emily home safe. Aaron introduced Mary to Emily's parents and briefly told them the story about the book. Don went from an emotional state to one of guarded excitement. "You're telling me that you kids uncovered a Russian plot to undermine the social structure of America?"

"Dad, it's true." Emily put her arm around Don. "And your friend Marty Smith was one of the prize students. The guy got all of his political savvy from one of the original perpetrators. I'll tell you the whole story over dinner. We're making a deal to exchange the book for some favors, and we can't divulge any of this information, so what I tell you goes no further."

Aaron kissed Emily goodbye, shook Don's hand, and hugged Trish. Mary hugged Emily and said goodbye to Don and Trish.

They stayed only a few minutes at Martha's and exchanged vehicles. Aaron promised to fill her in later. Martha was shaking her head as Aaron and Mary drove away. The law offices of Burkiwitz, Mellor, Bachman, and Jones were in one of the taller buildings in downtown Portland. The plush carpet looked new, and the waiting area was adorned with fancy J. Longstreet paintings and leather furniture. The receptionist sat behind a large computer desk. Aaron approached her and spoke. "We're here to see Mr. Burkiwitz. My name is Aaron Douglas."

"Oh yes, Mr. Douglas. Howard is expecting you." She pushed a button on her phone.

"Howard, Mr. Douglas and a young lady are here to see you."

"Please send them in." She pointed to a large oak door and told them to go in.

Howard Burkwitz sat behind a large oak desk. On one corner was a miniature Oakland Raiders helmet, and on the other corner was a miniature USC helmet. On the credenza behind him there was what appeared to be a game ball with a score written on it. It read "USC 31, Iowa 19."

"Did you play for USC?"

"I did, Aaron, about a hundred years ago. Who is your friend?"

"Mr. Burkiwitz, this is my sister, Mary." Howard stood and shook both Aaron's and Mary's hands.

"Sit down, and tell me what's on your mind."

It was after five when Aaron finished the story. Howard Burkiwitz looked to be almost in shock. He sat, pondering the situation for several minutes.

"So you want me to draft an amnesty agreement for the governor to sign, giving the people you've mentioned relief from any crimes they may have committed while serving as agents of Friends of the Forest? And any other acts that may have been against the law?"

Aaron thought for a minute. "Yes, I think that about covers it. What kind of a fee will we be charged?"

"I'll tell you what. If you wouldn't mind adding just one more name to your list, there will be no fee. As sweet as my Ann is, she has a sordid past with regards to environmental crimes against the state. I wouldn't mind seeing her have a clean slate."

Mary didn't wait for Aaron to answer. "That would be fine. By the way, how is she doing?"

"She's fine, but every now and then she has one hell of a nightmare."

Howard Burkiwitz promised the agreement would be a priority. He would have it ready the following Monday. Aaron and Mary left the law office and drove straight through to LaGrande.

It was after eleven when they walked into the house. As late as it was, Aaron decided to call Emily. Emily answered immediately.

"Hi, Emily," Aaron said. "Mary and I just got home safely. Everything is set; the document will be ready on Monday."

"Good. I can't wait to get this behind us. Do you realize we have to start school in less than three weeks?"

"Yes, I do, and I've never thought I'd be so excited to get back to school. We'll need to start looking for a place that splits the difference between Eugene and Corvallis. I'm sure we can find something before December." They talked until midnight, and Aaron promised to call her every day.

Aaron and Jed pulled into the mill yard just after four on Wednesday afternoon. A.J. was walking across the mill yard and yelled at Aaron. "Hey, what the hell did you do while we were gone?"

"Nothing out of the ordinary. Why?"

"Your grandma called me this morning and told me that you and Mary were in some kind of trouble." Jed patted Aaron on the shoulder and wished him luck. He then got into his truck and drove off. Aaron's mind was spinning a hundred miles an hour, trying to come up with a good story. It was no use; the truth was going to have to do. They went into A.J.'s office, and Aaron told A.J. everything. He made it clear that the information could go no further than the room they were in, and A.J. agreed. Aaron left nothing out, and A.J. sat there, totally stunned.

When Aaron finally finished the story, he sat and waited for the wrath of A.J. to come tumbling down around him. What he got instead was a handshake. "I've heard some crazy shit in my time, but never anything close to this. You guys actually went in this guy's house and found that book?"

"Oh, yes. We did everything I just told you about. It just seemed right at the time."

"The only thing that bothers me is the fact that there are people who committed some pretty serious crimes going scot-free."

"That bothers me too, but they are good people who were grossly misguided. And you know what?"

"What?"

"They won't do it again. Herb thinks that the FF is out of business. If just that is true, this whole mess was worth it."

"I guess you're right, Aaron. I guess you're right."

Aaron called Herb on his way home and asked him to set up a meeting with Marty on Tuesday. Herb wasn't able to get a hold of the governor until Thursday morning. They set a meeting up for 1:30 on Tuesday in the governor's office. Marty started to set out rules for the meeting, which Herb quickly trashed. "I will be there, as well as Aaron Douglas and our attorney."

"Whatever makes you happy. Just be sure that you have my property with you."

Herb couldn't resist. "You know, Governor, those books should be the property of the United States government. What we're doing is no noble act; it borders on treason."

"Save the bullshit for someone else." Marty slammed the phone down in his den just as Maggie was walking by. "What the hell is the matter with you? You know that really isn't your phone, don't you?"

"Maggie, shut up. Where were you last night?"

"Sleeping in one of the guest rooms."

"If you do that again, I'll come and get you. It won't be a pleasant experience. You shouldn't listen to everything that shrink tells you." Maggie left the room and walked out onto the backyard patio.

She stared off into the blue sky and thought to herself. "I don't know how, but some day I'll make you wish you'd never treated me the way you have. Some day, I'm going to get you back."

Chapter XXI

On Sunday morning, Aaron slept in until after seven. A.J. was putting on a pot of coffee as Aaron made his way to the kitchen.

"Morning, Son. Getting excited about your meeting?"

"I guess so. I want to get this whole thing behind us. Don't forget I'm getting married in just a couple of months."

"Have you gotten her a ring yet?"

"As a matter of fact I'm going to Boise today to pick one out."

"What's open in Boise?"

"Emily's aunt knows a jeweler who's very reasonable and has good stuff. She's going to help me pick one out."

"Have fun. I don't envy you. When are you giving it to her?"

"Tuesday afternoon, after the signing." Aaron was on the road by eight and was in Boise before eleven. Lynn met Aaron at the pizza-by-the-slice place where they had eaten lunch. They ate a quick lunch, and Aaron followed Lynn to a small shop just past the travel bureau.

The lady looked to be in her fifties and seemed extremely knowledgeable about the rings and other jewelry in her shop. Aaron looked at the selection, and in less than a half an hour, Lynn and he had picked out a beautiful three-quarter carat engagement ring with matching wedding band. It was about all Aaron could afford. "I'd like to get her something bigger, but I want to save enough to live on after we get married."

Lynn stood and put her hand on Aaron's shoulder. "Don't worry about bigger. This is a much better ring than most brides receive."

"Yes, but Emily isn't just any bride-to-be; she's Emily."

"You really love her a lot, don't you?"

"Yes, Lynn, I do. Thanks a lot for helping me. And thank you, Alice, for opening up on Sunday."

"Anything for a friend of Lynn's."

Aaron said goodbye and headed back to LaGrande. He called Emily just as he was leaving town.

"Hello?"

"Hi, Emily, guess what I just did?"

"What?"

"Can't tell you, but I'll show you on Tuesday."

"Tell me now."

"I love you, Emily." Aaron clicked the phone off and didn't answer when Emily called back. The sky was turning gray as Aaron pulled into a gas station in Baker City. He went into the convenience store and bought a bottle of root beer and a candy bar. Just as he was pulling out of the store, he saw the first flashes of lightning. The storm was huge, lasting until he was just east of LaGrande. All he could think about was the number of lightning strikes hitting the tinder-dry forest, especially the tinder-dry national forest.

He was home by 3:30 and found Mary and Dave out in the barn, saddling two horses. "Where are you guys going?"

Mary didn't answer the question. "Where is it?"

"Where's what, Mary?"

"The ring."

"Oh, in my truck. Want to take a look?"

"Of course I want to take a look! I'm a girl." They walked to the pickup, and Aaron opened the door and looked on the seat. The ring box was missing, and he felt his knees getting weak.

"It's gone! It was right here on the seat, and now it's gone."

"How could you lose it? Was it there when you got here?"

"I think so, but I can't really remember. I hope it didn't get stolen when I stopped for gas."

Mary saw the ring box on the floorboard and picked it up. "You want me to keep this until you need it?"

"Thanks. That scared the hell out of me! Take a look, and give me your honest opinion." Mary opened the box and looked at the ring. She showed it to Dave, and it was unanimous; the ring was perfect.

Mary and Nancy had done some preliminary planning, and they thought it was best to have one wedding and two receptions. Aaron liked the idea of two parties and only one wedding. Mary had been talking to Emily, and Emily to her mother, and everyone seemed to be on the same page. Aaron got down on one knee in front of Mary and proposed. "Emily, you are the light of my life, and I love you very much. Will you marry me?" Dave was laughing.

"I wish I'd had a recorder right then. Aaron Douglas proposes to his sister." Aaron stood up and looked at Dave.

"What's the matter, David? Did I make you jealous?"

Dave's face got beet red, and Mary saved him. "We better get going if we want to make it back before dark."

Aaron asked the question for a second time. "Where are you going?"

"I'm just going to show Dave around the ranch. Why?"

"Because if you're going up the hill, I'd suggest you take a rifle."

Aaron sat on the front porch and watched as they rode out of sight toward the west boundary of the ranch. He heard Nancy in the kitchen getting some pans out of the cupboard and thought to himself how odd it would be to live with a different person. He had no reservations about the wedding, or about how he felt about Emily, but still there was a feeling of uneasiness that he couldn't shake. He finally resigned himself to the fact that it had nothing to do with the wedding or Emily. It was all of the strange things that had happened over the summer. He wished there was a way the world could know, but he would never go back on his word to Emily, or the rest of the group.

Chapter XXII

Jonathon Finley sat back quietly in his first-class seat, listening to the jet engines getting ready to lift the big transcontinental plane off the ground. Amy Grayson sat in the adjoining seat and held the old professor's hand. The expedited passports had been a bargain at $120 apiece. And the one-way tickets to Rio would have been a bargain at ten times the price.

They had sold everything they had acquired over the past 48 years and stuck it in offshore accounts under their new names. Gary and Amanda Price had accumulated enough property and other worldly goods to amass a small fortune. They would live out their years drinking fine wine and smelling the warm Atlantic breezes drifting in over the small fishing village, just north of Rio. As the plane lifted off, Jonathon looked down at the lush Willamette Valley for the last time and thought to Himself, "What destruction I've done here. What devastation I've caused. I've helped put you in turmoil, and I'm not sorry. If it were not for the willing participants, I would have accomplished nothing. Thank you, you young venerable minds. Thank you, you senseless hordes of willing accomplices. Thank you, Marty Smith, my prize."

Marty sat behind the large oak desk going over budget papers, trying to figure out how to present a passable new tax. His medical plan and other major screw-ups had the state on the brink of bankruptcy. Marty didn't give a shit about the state's financial problems, or about budgets or anything else that pertained to his duties as governor. What Marty cared about was getting a hold of that revealing little book, getting through the next few months, and getting rid of Maggie. He had called Jonathon several times, wanting to tell him about his success in recovering the book, but he couldn't seem to track the old professor down.

Marty drove slowly down the country road and turned into the professor's driveway. There was a pickup truck backed up to the garage, and two men were loading what appeared to be some of the professor's belongings in the back. The shorter of the two recognized Marty.

"Governor Smith, if you're looking for Professor Finley, he has moved."

"Really? Where to?"

"I'm not sure, but I overheard someone say Europe."

"Did he sell this house?"

"Yes, the guy I work for bought this and a house near Coos Bay, paid cash for the both of them. He says he got one hell of a deal on them. They closed in one day. It was part of the deal."

"Thanks for the information."

At first, Marty was relieved that Jonathon was out of the picture. Then a sense of fear swept through him. Why was Finley running away? Did he plan to expose what he had done and whom he had so amply trained? Marty drove south on I-5 and took Highway 58 toward Oak Ridge. He drove to the burned ranger station and looked over the piles of charred remains. Marty sat there looking at the devastation, knowing that he had had a hand in the destruction. It was like he was in a movie, and yet he was a spectator. He thought of all of the immoral acts that he and the other Leaders of Freedom members had performed.

Marty wanted to distance himself from the carnage, but he just sat there in a paralyzed state. He kept telling himself to get over it and put the past behind him. At least Finley was out of the equation, and the manual would soon be in his hands. What could go wrong now? He was in the clear. He drove home and parked his car in the service garage. He put on a halfway decent face, thinking that he might get lucky if he were nice to Maggie.

Marty couldn't have been more wrong. The only sign of Maggie was a note, telling Marty that there was a special letter for him, and that she wouldn't be home until after ten. He walked from the entranceway into the den and picked up a funny-looking letter off his desk. The word *Screened* had been written across the letter, and the special code was in the proper place, telling Marty that the letter was safe. He walked into the kitchen and opened the door to the wine cellar. He grabbed a bottle of Merlot and used the corkscrew to open the bottle. He let the wine breathe while he carefully opened the letter.

The greeting sent chills down Marty's spine. It was from Professor Finley. Marty poured a large glass of wine into a goblet and carried it and the letter into his library. He sat on the overstuffed sofa in the corner, turned on the floor lamp, and began reading.

"Dear, Dear, Marty, I hope this letter finds you well. I myself am secluded in what can only be described as total euphoria. My years as a professor have certainly taken their toll and such a state is just what the doctor ordered. I wanted to take this time to further explain what went on during our little experiment. You see, Marty, I am writing you, and you alone, because you are my prize. I just wish all of my subjects could have excelled as you did. I knew that after the first week I met you, you were something special, and that you could help me. And oh, how you've helped me! My task was simple in theory, but in reality, it was extremely difficult to implement. Let me start by telling you what I set out to do.

I was to teach a totally different philosophy than what the students of the 1950's were accustomed to. I was to start slowly, so that no one would notice. I needed pupils whose minds could be transformed into what I needed. They were everywhere, just waiting for me. The people that got this crazy idea hit the nail right on the old head, Marty. They told us that if we were sincere and took our time, the very freedoms that made this country what it was could be used to severely damage it. I did start slowly, feeling out each student to see whether or not he or she was receptive. Then I culled the more resistant ones and paid little attention to them. The good ones, like you, Marty, were where I focused my time. I constantly fished to see how receptive to change each student was. Then, ever so slowly, I seduced each of you. You know how seduction works, don't you, Marty? What we wanted to do was to weaken a strong country, and by golly, Marty, we did just that. I'll be more specific. Let me explain what I did.

I trained students to become professors, lawyers, schoolteachers, and, of course, politicians. The politicians are my favorite. Do you know why they are my favorites, Marty? Let me tell you. They pick the judges, Marty, like so many that you've picked during your years in office. And in our great state, those judges are liberal judges. They are the best kind, aren't they, Marty? Because they don't really have much respect for the law, and even less respect for what is right and what is wrong. The thing I feared most were those nasty little bastards with common sense. I still get the shivers when I think about them. So what I did was help my students train other students and turn my little portion of the West Coast into a little haven for liberals. How did I do, Marty? Let's go over some of the finer points. The education system spends a great deal of money on each student, yet it yearns for more revenue. There may have been some bad ideas introduced. Did you help out with that, Marty? And isn't it great that the state controls all the land, and the people that buy the land have to beg and plead to improve it? But that's okay, Marty, because the state knows what's best for everyone."

Marty Smith threw the three-page letter on the floor, walked back into the kitchen, and poured another glass of the expensive Merlot. At the top of his lungs, Marty yelled, "Fuck you, Jonathon, you sorry son of a bitch!"

He carried the wineglass back into the library and scooped the letter off the floor. He almost tore the letter to shreds but curiosity got the better of him. He sat back down on the sofa and continued reading.

Social Manipulation

"And what about the medical plan you got started here? It is a beauty, isn't it? How's that credit rating, Marty? Can Oregon still borrow money if it needs to? I sure wouldn't want to pay those exorbitant rates. You know, Marty, with all of the loggers and mill workers we put out of business, there was a real need for free medical. You'd think all those high tech jobs that were going to fill the gap would provide better benefits, wouldn't you, Marty? What was it, Marty? 60,000 people that lost their jobs so that some old owl who would be just as happy in a Wal-Mart sign could have the woods all to itself? That's a hoot, isn't it, Marty? Better buy a good supply of hot dogs; there'll be plenty of fire.

I should just very lightly touch on the bigger picture. I'll be brief. The military is weakened every time a Democrat lands in the White House. Score one for us. Half of all Americans have a bleeding heart for a brain. That's two for the good guys, Marty. Are you keeping score? Let's see, unions have more power than the companies their members work for. That's not just companies, but government bodies as well. This just gets better and better, doesn't it, Marty? You'd think that the people with enough guts to put their money on the line would fare a little better, wouldn't you? Not in America, Marty. That's three for us. Oh, by the way, Marty, take those nasty guns away from those arrogant people. Where did they ever get the idea that they had the right to own guns? Oh, I could go on and on, but my plane leaves in just a few hours. I hope for your sake that you retrieve that nasty little book. It may incriminate you just a little.

Well, old friend, I guess this is it. Good luck during the remainder of your term. Oh, Marty, there is one more point I'd like to make before I sign off. I'm pretty sure that our little program was officially over about the same time the Berlin Wall came down. I continued to play the game because it was all I knew how to do. I often thought about quitting, but when you've lived something your whole miserable life, it's hard to give it up.

I must go now, but I'd like to leave you with this one troubling notion. For all of the effort that was put into this project, and all of the changes that were caused by it, the only thing it did was to make the

United States stronger...much, much stronger. I guess that's a point for the other side. Think about it. Marty. Your friend, Jonathon."

Marty stood and crumpled the letter in his hand. He walked into the large front room and turned on the gas fireplace. As he watched the letter burn, he thought to himself, "What the hell did he mean when he said it made us stronger?" Marty pondered the question until his head was spinning like a top. He went into the kitchen and poured another glass of wine. He sat in the front room and stared into the fire.

It was just before ten when Maggie walked through the front door. Marty stood and walked toward her. She had seen that look before and knew what the night was about to bring.

"Marty, I think you might have had too much wine. Why don't you go to bed and I'll be in there in a minute?"

"Oh, no, I don't think I want to go to bed. The evening is still young. Why don't you come here, and we can dance?" Maggie didn't move and stood, paralyzed, as Marty walked toward her. He put his arms around her and drove his lips against hers. There was no tenderness, no passion, and no emotion. She started to pull away, and he pushed her onto the sofa.

When he had finished, she sat, motionless, knowing that she had been raped again and knowing that there was nothing she could do about it. She felt dirty and ashamed, but all the time kept telling herself that it wasn't her fault. Marty finished off the bottle of wine and went to bed. She prayed for the day that the nightmare would end.

Chapter XXIII

Aaron drove straight to Emily's house without stopping. Trish answered the door while talking on the cordless and pointed to the stairs that led to the basement. He found Emily and Don engaged in a rousing game of foosball. Aaron waited for the point before he spoke. "Glad to see I'm not the only one." Don looked up.

"It's a girl thing; they cheat." Emily walked to Aaron and put her arms around him and kissed him. Aaron turned red; he had never kissed a girl right in front of her dad before.

Don spoke up. "Why don't I leave you kids alone?" He went up the stairs, and they kissed for what seemed like an hour.

"I'm glad December isn't that far away," Aaron said. "I don't think I could wait a whole lot longer."

"We don't have to wait. Let's do it right here, right now," she replied.

"Shhh. If your dad hears, you he'll kick my ass."

They ate lunch and then went to Robby's to get the books. They met Robby at the front door of his house, and he handed Aaron the bag.

"Thanks, Robby. I owe you one."

"Can you tell me what's going on?"

"I will some time, I promise." As Aaron drove toward Salem, Emily transferred the books from the bag to a plain paper sack.

Herb was waiting on the steps of the Capitol building and waved when he saw Aaron and Emily walking toward him. "You're a little early."

"Gee, Herb," Emily added, "you seemed to beat us here."

"I am a bit excited, Emily, and if things go right, we have our old lives back. I've done some crazy shit, and I'm not proud of any of it, but if I really get this second chance, I'm sure as hell going to take it."

They waited on the steps until Howard Burkiwitz met them. Howard questioned the fact that there was an extra person who was not scheduled. Herb looked at Aaron. "I think Emily is invited; after all, she was the one who was kidnapped." As soon as Marty's receptionist saw the group, she hit the call button and alerted Marty.

He didn't wait for them to enter his office. He met them in the hall and ushered them into his office. "Welcome. I see you all had a safe journey."

Emily couldn't resist the opening. "I almost didn't make it, Governor Smith. By the way, how's your goon doing? Last time I saw him, he was almost dead." Marty didn't respond, and Howard looked at Emily with a "Please shut up" look.

Howard laid the proposal out on Marty's desk, and Marty spent a good half an hour going over it. It was fairly simple. All of the people named in the document would be granted amnesty for any crimes that they might have committed prior to the date of the signing. Marty signed the document, and Howard witnessed the signing. There were two copies, and Howard took one and left one for Marty. Aaron took the paper bag he had been carrying and put it on Marty's desk.

"Here they are. It's the only ones in existence, to my knowledge. I am assuming that there were many more copies, but I'll bet they never show up." Marty took the books out of the bag and thumbed through the original until he found a page with his name written on it.

"Jesus, why did the old fart pick me?"

Emily couldn't resist the opportunity. "Easy pickins, I guess." Howard completed some unfinished business, asking Marty for the incriminating pictures of Emily pouring gas on the ranger station.

Marty handed him a package. "You don't have to worry about any federal charges stemming from the ranger station fire. It was ruled an accident. That was the only federal situation that came into play. The report is in this package. I think that is everything. Remember—this agreement is subject to no information being leaked by any of the recipients of the amnesty."

Howard spoke. "I think we're all on the same page. I haven't even looked at these books, and I don't ever want to, but by the sounds of the stories I've heard, I'd say they would have been very incriminating if they were exposed."

"The old man that owned the book told me that for all of the damage he had done, for all of the damage *they* had done, their only accomplishment was to make this nation stronger. What do you suppose he meant by that?"

Howard shrugged, shook the governor's hand, and motioned for the others to leave.

Once out of the building, Emily leaped into the air and yelled, "I'm free!" Aaron grabbed her and gave her a kiss.

"Emily, we're all free." Howard told them that he would have copies available by the end of the week, and that each person should secure a safe deposit box and keep the papers in them for at least 10 years. Herb assured him that they would all do so. Howard left, and the three walked to the parking lot. Aaron shook Herb's hand, and

Herb thanked Aaron for helping change his life. Emily, who wasn't always outwardly emotional, had tears coming down her face. Herb hugged her and walked toward his car.

"Hey, Herb," Aaron yelled. "See you at the wedding."

Herb turned around and spoke. "Oh, we'll see each other before that."

Aaron and Emily were back in Tigard before five, and Aaron was starving so they went to Emily's favorite pizza parlor and ordered her favorite. Aaron picked out a table that was in the far corner. They talked about how fortunate they were and how things had fallen into place. Out of nowhere, Aaron dropped to one knee and started to propose. "Emily, I had a speech prepared for this, but I forgot it. Will you marry me?" He opened the ring box and took the engagement ring out.

She offered her hand and answered. "I already said yes, but I'd love to say it again. Yes, Aaron, I'll marry you. This ring is beautiful." He stood up and put the ring on her finger and bent over and kissed her.

"I love you more than anything in the world, Emily."

"I love you more than anything in the world."

Chapter XXIV

Marty carried the paper bag into his den and laid it on the corner of his desk. He called Oswald and confirmed that the deal was complete. Marty was so excited that he didn't notice Maggie walking by the den entrance. She never eavesdropped on any of Marty's conversations, but the word "murder" had stopped her in her tracks. Marty had told Oswald that the incriminating books were in a paper bag on the corner of his desk. The word "incriminating" also caught Maggie's attention. The one-sided conversation led her to believe that Marty was implicated in the murder of Ernie Roper, and that the books in the paper bag on the corner of Marty's desk could possibly be used to destroy him. Through the reflection in the den window, she saw Marty hang up the phone and put the paper bag in his bottom right-hand desk drawer. She walked into the bedroom and lay on the big king-sized bed. How easy it would be to take those books and catch a plane to New York. Wouldn't Marty be surprised when he watched the evening news?

Marty walked to the wine cellar and took out what he believed was his finest bottle of Cabernet. As he uncorked the bottle, he saw Maggie walking through the front room toward him. Instead of her half-frightened look, she had a look of confidence on her face. Marty hadn't seen that look for years, and it excited him. She slowly walked over to him, grabbed two wineglasses off the rack and poured them half full. "Let's have a truce, Marty. I don't like the way we've been to each other. I can still make you feel like a man if you just give me a chance."

"Maggie, what in the hell's gotten into you?"

"Nothing. I just thought you might like it the way it used to be. You know, before you were really someone. Let's go sit on the deck, drink some wine, and see what happens."

He watched as she walked away, not taking his eyes from her butt. He grabbed his glass and the bottle and followed her to the deck. They drank wine and had a small dinner delivered. By nine, Marty had waited long enough. "Let's go to bed while I can still see."

"Come on, Marty. Where's your sense of adventure? Let's drink one more bottle. It was just after ten, and Marty, who had drank 90 percent of the three bottles, could hardly walk. She slowly undressed and threw her clothes on the floor. Marty hadn't realized just how sexy Maggie could be. He hadn't seen her act like this in years. When she had stripped to her bra and panties, she pulled the covers back and climbed in between the satin sheets. Marty tore at his own clothes and as soon as he was completely undressed, he lay down beside her and began touching her all over. Maggie did nothing to stimulate Marty, and the harder he tried to achieve an erection, the more out of reach it was. Finally, Marty gave up and drifted off into a deep sleep.

By midnight, Maggie was on the road to Seattle. She had tried Portland, but the first flight to New York didn't leave until 8 A.M. There was a 6 A.M. flight out of SeaTac, and Maggie had a reservation on it. It was the only direct flight from the Northwest. She hadn't taken the time to look at the books she had taken from Marty's desk. If they wouldn't ruin Marty like she hoped they would, she herself would be ruined. She would have time to look at the books as soon as she was safely en route.

It was just after nine, and the phone by Marty's bed wouldn't stop ringing. "Maggie, would you please get that?" The ringing stopped just as Marty picked up the receiver, but the caller had already hung up.

Marty's head was pounding, and his vision was blurred. He put on a robe and wandered through the mansion, looking for Maggie. When he looked in the service garage and found that her car was missing, he figured that she had gone for coffee. He hoped that she would bring him an extra large cup. He went into his office and sat behind his desk. He had a meeting at 11 A.M. and was relieved to see that it was only 9:30.

He opened the bottom desk drawer and was relieved to see the paper bag. Before his meeting, he would burn the two books and forever hide the incriminating evidence.

When the plane was two hours out of New York, Maggie took the books from her carry-on bag and started thumbing through the translated copy. She couldn't believe what she was reading. It was so much better than she had thought it would be. She went over and over the part about picking the perfect students to achieve the specific goals of getting the chosen subjects into government. Marty's name appeared several times on the edges of some of the pages. Maggie looked at the original to make sure that the names were also written in it. The writing looked old, and she knew that the book was authentic. There was only one network that she would trust with the books. She would take them to ANN.

American News Network was the only one she could trust. She knew the big three would spin the story and even if they aired it, they would do everything they could to discredit it. She wasn't worried about Marty finding her. He would never believe that she would do something like this to him. The horrible years she had spent with him were about to end. Whatever the outcome, there was no Marty in her future.

Marty took an extra long shower and put on one of his finest suits. He took four pain pills and washed them down with some warm tap water. He walked to the den and took the paper bag from his desk drawer. He walked into the front room and turned on the gas fireplace. He would burn the pages a few at a time, and then rejoice as he watched the bindings slowly melt away in a cloud of blue smoke.

Marty took the first of the two books out of the bag and stood, staring at the cover of the cheap paperback novel. The title read *A Knife in the Back*. He took the second book from the bag and began to shake violently as he read its title: *The Last Rape of Satin*. Marty looked at the books and began cursing Maggie. What kind of game was she playing? She would pay dearly for this little escapade. Marty phoned Oswald to tell him of their newest problem. "You need to find her, and find her fast. I don't know what that crazy bitch is capable of, and I don't want to find out."

"Any idea where she might have gone?"

"No, but you might try her mother's. That's been one of her sanctuaries in the past."

"Don't worry. I'll find her." Oswald couldn't help putting a little salt into Marty's wound. "I just hope it's not too late."

" Christ's sake, Oswald, don't talk like that. Now go find her!" He slammed the phone into Oswald's ear and kicked the sofa as hard as he could, hurting his big toe. Oswald hung the up phone, walked out of his office and drove home. He hooked up his boat and drove to his favorite fishing spot.

It was just after three when Maggie walked into the terminal at JFK International. She had no luggage to retrieve, so she went directly to the taxi area by arriving flights. She took the first taxi and told the driver to take her to ANN headquarters. He made his way through the crowded streets, and in less than half an hour she was in the lobby of the most trusted news network in America. The receptionist was in her thirties and was aptly schooled in the art of gate keeping.

"May I help you?"

"Yes, perhaps you may. I have with me one of the biggest stories of the century, and I am willing to give it to your network, if they will release the information."

"I'm sorry, but you will have to call ahead for an appointment. There just isn't anyone available to see you right now."

"Listen here. You forget all that shit they told you about getting rid of crackpots. You get on that phone and call someone with authority and get him or her down here now. If you don't do that, I'll take this to another network and make damn sure these people here know why they didn't get a first look at the story."

The receptionist hit three buttons on her phone. "Sir, there is a lady here that I think someone should talk with. She has what she described as one of the biggest stories of the century." She clicked off her phone and looked at Maggie. "Mr. Conners will be right down. I'm sorry, but we get a lot of people in here claiming to have big stories for us. It's hard to sort out the serious ones."

"You did the right thing this time," Maggie assured her.

Marty missed the 11 A.M. meeting and spent the better part of the day trying to get a hold of Oswald. He called Maggie's mother, and there was no one there. Marty cursed himself for not destroying the books when he had had the chance. He was going to do it last night, but that manipulating bitch had distracted him. He would seize her by the throat and not let go until she had taken her last disgusting breath. He kept telling himself that there was nothing to worry about. She wouldn't do anything to disgrace the highest office in the state. What had he done to her that would make her do such a horrible act? All of those episodes in the bedroom were natural, just a game that they played. All of the crying afterwards was just part of the game.

Jack Conners was a tall man who looked to be in his mid-forties. He approached Maggie and extended his hand.

"Hello, I'm Jack Conners. I understand you have some information that we may be interested in."

"Mr. Conners, may I ask what it is you do here at ANN?"

"I'm the vice-president in charge of information procurement. You have the right person." Maggie extended her hand and shook his.

"I'm Maggie Smith. My husband, Marty, is the governor of Oregon." Maggie could see an instant change in Jack Conners' attitude.

"Maggie, what can we do for you?"

"Mr. Conners, I don't mean to sound rude, but I would prefer to discuss this with at least one more person present."

"That could be arranged. Would Matt Tracy be an adequate addition?" Maggie thought he was kidding.

"Sure, I think Matt Tracy would do just fine." Matt Tracy was the anchor for the six o'clock edition of ANN News Live. He was the most watched news broadcaster in the world. Jack Conners walked toward the elevators.

"Follow me, Maggie." They grabbed an elevator and went up several stories. The elevator stopped, and they exited onto a plush hallway. Jack Conners knocked on the door with Matt Tracy's name on the nameplate. Before he heard an answer, he opened the door and waited for Maggie to enter the office.

Matt stood and introduced himself. "Hi, I'm Matt Tracy. You must have some pretty good stuff."

Jack spoke up. "Matt, I've not seen the information yet. This is Maggie Smith. Her husband is Marty Smith, the governor of Oregon."

Maggie sat in one of the plush chairs and slowly took the books out of her bag. She handed both books to Matt. "I think you'll find theses books quite interesting, Mr. Tracy."

"Please, Maggie; call me Matt."

"Okay, Matt. I'll tell you what I know about these. The original is a Russian manual that was used by a professor at the U of O. He was on staff there from 1953 through 1999. I haven't read the entire translated copy, but what I have read is utterly amazing."

It was after four when Matt started looking through the translated copy. At five, he closed the book and looked at Maggie. "You do know that this won't do your husband's career any good?"

"I'm sure it won't, but the fact that he was in on the murder of a potential informant won't do it any good, either."

Matt Tracy stood up and put his hands over his ears. "Did I hear the word 'murder'? Please tell me I didn't hear the word 'murder'."

"I did say 'murder', and here is a list of some of my husband's assistants who might be able to shed some light on the crime. As you by now realize, I don't give a shit about what happens to Marty Smith. If he goes to jail for 20 years, that will suit me just fine."

Jack Conners, who had been witnessing the conversation, spoke up. "Matt, did you ever hear about that couple, I think they were from Iowa or somewhere in the Midwest, who claimed to be Russian spies, sent here to have their kids raised as American professors? I think the story was in some ancient tabloid. It was always shelved as a hoax, but this fits right in with the story."

Matt looked at the books and then at Jack. "Jack, find out all you can about that story, and have our best two investigative reporters on the next plane to Oregon." He looked at Maggie. "What was the name of the murder victim?"

"It's right there on the top of the page, with the 'V' behind it. Ernie Roper was his name. My husband was involved in some kind of a secret organization. I only got bits and pieces from time to time. I do know that it had something to do with environmental type issues."

"Maggie, I go on the air shortly. Can we keep these books and verify some of the information before we air it?"

"I think that would be wise."

"Maggie, if you have time, I'd like you to join me for dinner. We'll put you up in one of the finer hotels, and I'll have a limo pick you up around seven."

"That would be fine, Matt. It will give me time to do a little shopping."

Chapter XXV

Emily's last day at the swimming pool was Wednesday, and early Thursday she was on the road to LaGrande. Nancy, Mary and she would plan the eastern Oregon wedding reception. School would start in just two weeks, and Emily wanted all the planning completed prior to that. She felt so relieved about the amnesty agreement. Not so much for herself, but for Rachael. It was just fate that had brought Aaron and her together; for that, she was thankful for her short stint in FF. Emily and Trish had everything ready for the reception in Portland. They would be married on the first Friday of Christmas vacation and have the Portland reception that evening. Then on Saturday evening, they would have the reception in LaGrande. On Sunday, they would leave for their honeymoon and the start of a new life together.

Aaron had talked A.J. into letting him have the rest of the week off. At noon, he left for the ranch. He was sitting on the front porch when Emily pulled into the driveway. Aaron had never seen Emily run, and he laughed when she sprinted from her Explorer to the front porch. "You're actually pretty graceful when you run."

"Thanks. I always try to be graceful." Aaron hugged Emily and kissed her for several seconds.

"So," asked Emily, "what's on tap after the wedding planning is over?"

"How about another trip up to the lake, and we'll see if there are any of those monster rainbow trout left?"

"That sounds like fun. Only one thing I'd ask."

"And what might that be?"

"That we take a rifle with us." Aaron assured Emily that they would be amply armed.

Emily, Mary, and Nancy had all of the plans arranged by two, and Aaron had the horses saddled and ready to go. They were at the lake by 3:30, and Emily had her line in the water by the time Aaron had the horses unsaddled and tied up. The fishing wasn't as good as it had been in late spring, but in a span of two hours, they had enough fish for dinner. Aaron cleaned the fish while Emily saddled the horses.

Emily walked over to the edge of the lake and put her hands on Aaron's shoulders. "Are you mad about not being able to tell the world about Finley and Marty Smith?"

"I'm sorry that they're getting away with murder."

It was business as usual for Marty Smith. He had stopped worrying about Maggie. He was sure that she would come begging back, with the books in hand. It was Oswald that worried Marty. Marty had been trying to track Oswald down for the better part of two days. He had no idea that two reporters from the American News Network were interviewing Oswald. He had no idea that he was being implicated in the murder of Ernie Roper. He had no idea that Maggie was about to get her wish.

Maggie slept in until after ten. The dinner with Matt Tracy had lasted well into the night. He was so grateful that she had given ANN the chance to break the story. She showered and got ready for what was to be the happiest day of her life. She would have lunch with Mr. Tracy, do a little shopping, and be at the ANN studios just before the story was aired. She had no remorse for what she was doing. The horrible years that she had spent as a prisoner in the governor's mansion were more than enough reason. Her years of suffering were about to end.

It was just after six when the phone at the ranch rang. Mary answered the call from Rachael.

"Is Emily there?"

"Is this Rachael?"

"Yes. How are you, Mary? Are you watching it?"

"Watching what?"

"Oh God, Mary, turn on ANN. You are not going to believe it." Mary hung the up phone and turned on the big screen in the front room.

The caption scrolling across the bottom of the screen said it all. "Story of the Century." There were pictures of the governor's mansion with state police cars in front. Matt Tracy was telling the world of the Russian plot that had been altering the American way of life for so many years.

A.J. and Nancy, who had just walked into the house, stood with their eyes glued to the screen. A.J. finally spoke. "Where are Aaron and Emily?"

Mary answered, "They rode up to the lake."

They rode over the last rise before the trail dropped down to the ranch. "Aaron, what do you think Finley meant when he said that the whole plot made us stronger?"

"I have a theory about that."

"Like the one you had about the old guys who preach abstinence?"

"Yeah, kinda like that. I think that what they preached was wrong, and by pushing so hard against the truth, the truth just keeps getting stronger and stronger. Do you like that one?"

"I like it, Aaron."

The End

Printed in the United States
16133LVS00004B/229-357